Nucleation of Particles from the Gas Phase

The formation of small solid and liquid particles is vital for a variety of natural and technological phenomena, from the evolution of the universe through atmospheric air pollution and global climate change. Despite its importance, nucleation is still not well understood, and this unique book addresses that need. It develops the theory of nucleation from first principles in a comprehensive and clear way and uniquely brings together classical theory with contemporary atomistic approaches. Important real-world situations are considered, and insight is given into cases typically not considered such as particle formation in flames and plasmas. Written by an author with more than 35 years of experience in the field, this will be an invaluable reference for senior undergraduates and graduate students in a number of disciplines, as well as for researchers in fields ranging from climate science and astrophysics to the design of systems for semiconductor processing and materials synthesis.

Steven L. Girshick is Professor Emeritus in the Department of Mechanical Engineering at the University of Minnesota. He served as editor-in-chief of the journal *Plasma Chemistry and Plasma Processing* and was the founding president of the International Plasma Chemistry Society. His awards include the Boit Prize in Writing at MIT, the 2005 Plasma Chemistry Award, and the University of Minnesota Best Director of Graduate Studies Award.

"This book on particle nucleation from the gas phase is remarkably valuable as both an introduction to the field as well as a succinct summary of the current literature. The treatment strikes an appropriate balance between rigor and readability. Highly recommended."

David Graves, Princeton University

"This book provides a comprehensive foundation across disciplines, serving as the definitive guide for students and researchers alike. Offering insights into both classical theories and modern approaches, it promises to be an indispensable resource, fostering connections and advancing knowledge in a critical yet underexplored field."

A. M. Boies, University of Cambridge

"A concise and highly readable treatise on gas-phase nucleation, useful for graduate students and those interested in developing a deeper understanding."

Michael Zachariah, University of California, Riverside

"An essential text for scientists and engineers looking to understand the fundamental theories describing homogeneous nucleation, the state-of-the-art atomistic models, and more complex multicomponent routes to nucleation. This book is also an excellent reference text, as each chapter briefly reviews prior chapters, and introduces material in a self-contained manner."

Chris Hogan, University of Minnesota

"An exceptionally clear introduction to the theory of nucleation from the gas phase, this book provides a much-needed unified treatment of a broad range of nucleation phenomena. It is of great value to scientists and engineers in fields from atmospheric science to nanomaterial synthesis being both rigorous and accessible."

Mark Swihart, University at Buffalo

"A must read for all those interested in understanding the intricacies of nucleation. In this remarkably accessible book, readers from advanced undergraduate students to senior researchers will learn about the current state of the science."

George Shields, Furman University

"Written by a well-known expert who has particularly contributed to the issue of self-consistent classical theory, this book provides a nice overview. This book will prove useful for several current research areas, such as atmospheric science and global climate change."

Paul Wagner, University of Vienna

"This is the book I have been looking for! From the basics of nucleation theories to the applied aspects of chemical nucleation and nucleation in plasmas, it is easy to read and provides end-of-chapter problems. This is the ideal book for students and engineers."

Pere Roca i Cabarrocas, CNRS, Institut Polytechnique de Paris

Nucleation of Particles from the Gas Phase

STEVEN L. GIRSHICK

University of Minnesota

CAMBRIDGE
UNIVERSITY PRESS

Shaftesbury Road, Cambridge CB2 8EA, United Kingdom

One Liberty Plaza, 20th Floor, New York, NY 10006, USA

477 Williamstown Road, Port Melbourne, VIC 3207, Australia

314–321, 3rd Floor, Plot 3, Splendor Forum, Jasola District Centre, New Delhi – 110025, India

103 Penang Road, #05-06/07, Visioncrest Commercial, Singapore 238467

Cambridge University Press is part of Cambridge University Press & Assessment,
a department of the University of Cambridge.

We share the University's mission to contribute to society through the pursuit of
education, learning and research at the highest international levels of excellence.

www.cambridge.org
Information on this title: www.cambridge.org/9780521820530
DOI: 10.1017/9781139028851

First published 2024

A catalogue record for this publication is available from the British Library

Library of Congress Cataloging-in-Publication Data
Names: Girshick, Steven L., author.
Title: Nucleation of particles from the gas phase / Steven L. Girshick.
Description: Cambridge ; New York, NY : Cambridge University Press, 2024. | Includes bibliographical
    references and index.
Identifiers: LCCN 2024000794 (print) | LCCN 2024000795 (ebook) | ISBN 9780521820530 (hardback) |
    ISBN 9781139028851 (ebook)
Subjects: LCSH: Nucleation. | Gases–Liquefaction. | Kinetic theory of gases.
Classification: LCC TP156.N8 G57 2024 (print) | LCC TP156.N8 (ebook) | DDC 660/.043–dc23/eng/20240314
LC record available at https://lccn.loc.gov/2024000794
LC ebook record available at https://lccn.loc.gov/2024000795

ISBN 978-0-521-82053-0 Hardback

To Cyrus, Samaya, and Kieran

Contents

Preface

This book is concerned with the earliest stages of the formation of condensed-phase particles from molecules in a gas. This process, known as "gas-phase nucleation," is of considerable importance in a vast range of natural and technological phenomena, such as air pollution and global climate change, formation of interstellar dust, soot formation in combustion processes, plasma processing of semiconductors, engineered synthesis of nanoparticles for a wide variety of applications, and others.

Considering the importance of gas-phase nucleation to so many fields, it is remarkable that it remains so poorly understood. No theory presently exists, aside from empirical fits to the data for a given substance, that can predict the gas-phase nucleation rate of a given substance under given conditions, with anything approaching reasonable quantitative accuracy when compared to experiments, or even the qualitative dependence of the nucleation rate on conditions, except in the crudest approximate way. This situation, together with the paucity of literature that introduces the subject to graduate students or researchers in fields where gas-phase nucleation is relevant, but who are not specialists on the topic, has motivated me to write this book.

For almost a century, a considerable body of technical literature has grown around the topic. Overwhelmingly, this literature has focused on the simplest scenarios involving the condensation of supersaturated vapors to form particles. Yet many real-world situations, such as soot formation in hydrocarbon combustion or nanoparticle formation in plasmas, do not follow these simple scenarios, and the growing literature on each of those subjects has been self-contained, having seemingly little in common with the conventional understanding of gas-phase nucleation. It is true that these different scenarios require different concepts and approaches to understanding particle formation. Nevertheless, it seems surprising that almost no literature exists that brings together these seemingly disparate situations. After all, each involves the formation of condensed-phase particles from gases where no preexisting particles or bounding surfaces exist, so it would seem that they have much in common.

I hope that this book at least begins to rectify this situation, by putting under one cover an introduction to classical nucleation theory, including single-component, multi-component, and ion-induced nucleation, together with coverage of contemporary atomistic approaches, transient nucleation, chemical nucleation, and nucleation in plasmas.

I have endeavored to make this book comprehensible to students who have a good undergraduate background in some field of the physical sciences or engineering. In particular, it is assumed that students are familiar with undergraduate thermodynamics, as commonly taught in various undergraduate science and engineering disciplines.

This book grew out of a number of papers in technical journals that I published as author or coauthor, as well as from lecture notes I developed for graduate courses at the University of Minnesota Department of Mechanical Engineering that included a five-week "advanced topics" course on "Nucleation of Particles from the Gas Phase," a course on "Advanced Aerosol/Particle Engineering," and an interdisciplinary course titled "Introduction to Nanoparticle Science and Engineering" that brought together faculty and students from several different science and engineering departments. Similar courses are increasingly offered in a number of graduate programs in the United States and other countries, with several different academic disciplines represented. Instructors and students in such courses should find this book useful, as should researchers in the field of nucleation itself, or in the wide variety of fields where gas-phase nucleation is an important phenomenon.

I am pleased to acknowledge the contributions to the work presented in this book that were made by my Minnesota faculty colleagues, former postdoctoral associates, and PhD students. These include Peter McMurry and Chia-Pin Chiu, on self-consistent classical nucleation theory; Nagaraja Rao, on ion-induced nucleation; Donald Truhlar, on atomistic approaches to nucleation; Michael Zachariah and Song-Moon Suh, on transient nucleation; Mark Swihart, on chemical nucleation; and Uwe Kortshagen, Pulkit Agarwal, Upendra Bhandarkar, Sarah Warthesen Desotell, and Romain Le Picard, on nucleation in plasmas. I have been fortunate indeed to have such a distinguished and collegial group of colleagues and students with whom to work.

Symbols

$a_{A,l}$	activity of substance A in the liquid phase of a multicomponent solution, defined by equation (4.18)
$a_{A,v}$	activity of substance A in the vapor phase, defined by equation (4.17); identical to saturation ratio S
A	generic chemical species
A	factor by which the need for third-body stabilization affects the dimerization rate
A_n	cluster of substance A containing n monomers
A_n^*	excited complex or transition state in a chemical reaction
A_nB_m	binary cluster consisting of n molecules of substance A and m molecules of substance B
A_1B_i	initiating species in a sequence of reactions given by reaction (8.11)
A_1B_j	growth species in a sequence of reactions given by reaction (8.11)
ane_m	a silane containing m silicon atoms
B_k	by-product species in a sequence of reactions given by reaction (8.11)
c_v	specific heat at constant volume ($J \cdot mol^{-1} \cdot K^{-1}$)
c_p	specific heat at constant pressure ($J \cdot mol^{-1} \cdot K^{-1}$)
$\mathbf{c}$	velocity ($m \cdot s^{-1}$) of cluster or particle due to all effects other than convection and diffusion
C	capacitance (F)
C_0	capacitance of vacuum (F)
$\overline{C}$	mean thermal speed of molecules in a gas ($m \cdot s^{-1}$)
d	diameter (m)
D	diffusion coefficient ($m^2 \cdot s^{-1}$)
e	elementary charge, 1.602×10^{-19} C
E	energy (J)
$\mathbf{E}$	electric field ($V \cdot m^{-1}$ or $N \cdot C^{-1}$)
$\mathbf{E}_0$	electric field around point charge in vacuum ($V \cdot m^{-1}$ or $N \cdot C^{-1}$)
EA	electron affinity (J or eV)
f_{int}	fraction of atoms in the interior of a cluster
f^*	rate of monomer addition to a single critical cluster (s^{-1})
G	Gibbs free energy (J)
ΔG^0	change in Gibbs free energy (J) for a reaction, evaluated at standard pressure
$\Delta G_{n-1,n}^0$	stepwise change in standard Gibbs free energy (J)
ΔG_n	Gibbs free energy of formation of an n-mer from the monomer vapor (J)

ΔG^*	value of ΔG_n evaluated at $n = n^*$
$\Delta_f G$	Gibbs free energy of formation of a chemical substance from its standard reference elements (J)
H	enthalpy (J)
H	dimensionless function defined by equations (3.71) and (3.72)
$\Delta_f H^0$	standard enthalpy of formation of a chemical species (J)
I	generic ion
I$\cdot$0	same as I, emphasizing that the ion has zero attached monomers
I$\cdot$A$_n$	an ionic n-mer, consisting of an ion plus n monomers of substance A
I_{LIPEE}	signal intensity in laser-induced particle explosive evaporation
J	nucleation rate (m$^{-3}\cdot$s^{-1})
J_n	nucleation current for clusters of size n (m$^{-3}\cdot$s^{-1})
J_n^*	dimensionless nucleation current, defined by equation (7.14)
J_0	pre-exponential term in steady-state nucleation rate in equation (3.91)
k	rate constant (units vary)
k_B	Boltzmann constant (1.381×10^{-23} J$\cdot$K^{-1})
k_L	Langevin rate constant (m$^3\cdot$s^{-1}) for ion-neutral collisions, given by equation (9.8)
k_n	rate constant (m$^3\cdot$s^{-1}) for the forward direction of reaction (2.9); the condensation rate constant
k_{-n}	rate constant (s^{-1}) for the reverse direction of reaction (2.9); the evaporation rate constant
$k_{n(M)}$	rate constant (m$^6\cdot$s^{-1}) for reaction (6.7)
k_n^*	rate constant (m$^3\cdot$s^{-1}) for reaction (6.8)
K	dimensionless equilibrium constant for a reaction
K_c	concentration equilibrium constant for a reaction
$K_{n-1,n}$	dimensionless equilibrium constant for reaction (2.9)
L	heat of vaporization (latent heat) (J$\cdot$mol^{-1})
m	mass (kg)
m	number of B molecules in a binary A-B cluster
m_{AB}	reduced mass of A-B collision partners (kg), defined by equation (3.40)
m_r	reduced mass of ion-neutral collision partners in equation (9.8), in amu
M	third body in a three-body molecular collision
M	an arbitrarily large integer
$\widehat{M}$	molar mass (molecular weight) (g$\cdot$mol^{-1})
n	number of molecules in a cluster
n	number of A molecules in a binary A-B cluster
n^*	number of molecules in a critical cluster
$n_{\exp}^*$	inferred experimental value of the number of molecules in a critical cluster, based on equation (2.86)
n_0	in ion-induced nucleation, the number of monomers in a stable prenucleus
n_0	maximum cluster size for which atomistic data are available
N	number density (m^{-3})
N	width in number of atoms of a cube-shaped cluster

N^*	dimensionless number of stable particles, defined by equation (3.128)
N_n	n-mer number density
N_n^*	dimensionless n-mer number density, defined by equation (7.15)
N^0	standard number density, defined by equation (2.19)
N_A	Avogadro constant (6.022×10^{23} mol^{-1})
N_e	number density of free electrons (m^{-3})
N_p	particle number density (m^{-3})
N_+	number density of positive ions (m^{-3})
N_-	number density of negative ions (m^{-3})
N$_{tot}$	total number of molecules in a system
p	pressure (Pa)
p_A	partial pressure of substance A (Pa)
p^0	standard pressure, usually 1 atm or 1 bar
$p_{i,eq}$	partial pressure of a vapor of substance i existing in equilibrium with the flat surface of a bulk multicomponent solution having the same composition and at the same temperature as a multicomponent cluster (Pa)
q	electric charge (C)
r	radius (m)
r_i	radius of ionic core in Thomson model of an ionic cluster (m)
r_n	radius of ionic droplet (I·A$_n$) in Thomson model (m)
$\mathbf{r}$	spatial location relative to the origin of a coordinate system
R	molar gas constant, 8.314 J·mol^{-1}·K^{-1}
R	dimensionless cluster radius, defined by equation (5.40)
R_n	reaction rate (m^{-3}·s^{-1}) for the reaction that produces an n-mer by the condensation of one monomer
R_{-n}	reaction rate (m^{-3}·s^{-1}) for the reaction in which one monomer evaporates from an n-mer
s	surface area (m^2)
S	saturation ratio
S	entropy (J·mol^{-1}·K^{-1})
S_i'	in multicomponent nucleation, the effective supersaturation of a vapor component i, defined by equation (4.6)
Sipdr	silicon powder
syl$_n$	silylene containing n silicon atoms
t	time (s)
t^*	dimensionless time, defined by equation (7.16)
T	absolute temperature (K)
$\mathbf{u}$	bulk fluid velocity (m·s^{-1})
U	electrical energy stored in a capacitor (J)
v	volume of a cluster (m^3)
v_1	molecular volume (m^3)
V	volume of region of space (m^3)
W	work (J)

$W_{E,n}$	work against the electric force required to remove n monomers from an ionic droplet
x	number of molecules in a cluster, treated as a continuous variable
x	distance from origin in x-direction of Cartesian coordinate system (m)
x^*	number of molecules in a critical cluster, treated as a continuous variable, determined by equation (3.31)
y	distance from origin in y-direction of Cartesian coordinate system (m)
z	distance from origin in z-direction of Cartesian coordinate system (m)
$z_{p(-)}$	average negative charge per particle
Z	bimolecular collision rate (m^{-3}·s^{-1})
Z	Zeldovich factor
α	sticking coefficient
α	ratio of number density of growth species to that of by-product species in chemical nucleation, defined by equation (8.23)
α	polarizability of neutral molecule in equation (9.8), in Å^3
β	monomer flux (m^{-2}·s^{-1})
$\beta_{A,B}$	collision frequency function (m^3·s^{-1}) for collisions between molecules of species A and B, defined by the term in parentheses in equation (3.44)
χ	mole fraction
δ	infinitesimal quantity
δ	Tolman length (m), defined by equation (3.61)
$\delta_{2,n}$	Kronecker delta function for monomer evaporation from a dimer, defined by equation (3.127)
ε	relative permittivity (dielectric constant)
ε_0	vacuum permittivity, 8.85×10^{-12} F·m^{-1}
ϕ	electric potential (V)
Φ	dimensionless electrical work defined by equation (5.38)
γ	ratio of specific heats, c_p/c_v
Γ	diffusive flux (m^{-2}·s^{-1})
λ	mean free path (m)
μ	chemical potential (J)
ρ	mass density (kg·m^{-3})
ρ^c	net charge density (C·m^{-3})
σ	surface tension (N·m^{-1})
σ_0	surface tension of a flat liquid surface in equilibrium with its vapor
τ	characteristic time (s)
τ	dimensionless time, defined by equation (3.129)
τ^*	dimensionless time lag required to reach steady-state nucleation, based on equation (7.24)
Θ	dimensionless surface tension, defined by equation (3.28)
$\dot{\omega}$	net volumetric rate of species generation by reactions (m^{-3}·s^{-1})
Ω	collision cross section (m^2)
ζ	a constant in equation (3.37) that equals either 1 or 2

Subscripts

A	neutral monomer of substance A
CNT	classical nucleation theory
e	electrons
eq	at an equilibrium state
exp	experimental
i	ion
l	liquid
m	number of B monomers in a binary A-B cluster
n	number of monomers in a cluster, or of A monomers in a binary A-B cluster
$n-1, n$	pertaining to the reaction in which a cluster grows from size $n-1$ to size n
p	condensed-phase particles
rev	reversible
s	at an equilibrium saturation state
SS	steady state
v	vapor
x	associated with continuous size variable
x	component of a vector in the x-direction
y	component of a vector in the y-direction
z	component of a vector in the z-direction
0	vacuum
−	negatively charged
+	positively charged

Superscripts

z	number of charges on an ion
0	evaluated at standard reference pressure
^	molar quantity; per mole
~	per molecule
*	associated with critical nucleus
*	denotes value of saturation ratio above which nucleation lies in collision-controlled regime

1 Introduction

Nucleation is the birth of a new phase of matter from an old phase. This occurs when molecules, in sufficient number, configure themselves so as to constitute a "nucleus" of the new phase. The *Oxford English Dictionary* defines "nucleation" as "the formation of nuclei, esp. by the aggregation of molecules into a new phase within a medium." The Latin word "nucleus" means "kernel" and is related to the English word "nut" (in Latin "nucula" or "nux") (Simpson and Weiner 1989).

1.1 Significance of Topic

This text is concerned with the nucleation of condensed-phase particles, either solid or liquid, from the gas phase. This process is important for a wide variety of natural and technological phenomena. For example, obtaining a better understanding of the nucleation of particles in the atmosphere is necessary for increasing the accuracy of models for predicting global climate change. Soot formation during combustion is of central importance for fields such as air pollution, design of power plants and combustion engines, and fire science. The nucleation of interstellar dust is thought to play a crucial role in the evolution of the universe. In nuclear fusion reactors, dust formation in the edge plasma adjacent to reactor walls poses a critical challenge to the successful development of fusion as a global energy technology. As characteristic feature sizes in microelectronics shrink to smaller and smaller dimensions, particles that can form by nucleation in the chemically reactive plasmas used for semiconductor processing pose a serious contamination threat that can lead to "killer defects" in integrated circuits. Conversely, deliberate nucleation from the gas phase is one of the main routes to generating nanoparticles for use in a broad range of advanced technologies, including, for example, advanced ceramics, superhard coatings, catalysts, improved solar cells, xerography, ultraviolet-blocking cosmetics, medical diagnostics and therapies, magnetic materials, and electronic devices. Nucleation of particles from the gas phase creates an "aerosol" – a population of particles that are dispersed in a gas – and aerosol science is itself an important discipline with many applications.

An understanding of particle nucleation from the gas phase is thus of practical interest to scientists and engineers in a wide variety of disciplines.

It is also of fundamental interest, as nucleation represents one of the major unsolved problems of science. Despite attracting the interest of many scientists, who over the past century have produced a large body of literature on the subject, it is still not

1

possible to predict, with reasonable quantitative accuracy, some of the most basic features of nucleation, such as the *nucleation rate* (the rate of formation of new nuclei per unit volume) and the dependence of the nucleation rate on temperature and vapor concentrations. This statement applies even for the simplest nucleation scenario – homogeneous nucleation of a single chemical substance – for the vast majority of chemical substances. That this is so has much to do with the fact that the science of atomic and molecular *clusters* – including the "aggregation of molecules" comprising a nucleus – now the focus of great interest, is still in its infancy.

1.2   Types and Regimes of Nucleation

1.2.1   Nucleation in General

We know from everyday experience that matter can exist in various phases – solid, liquid, or gas – and that the phase of a given substance depends on its temperature and pressure. Thermodynamics teaches us that the phases of a given chemical substance at equilibrium can in general be represented on a diagram such as Figure 1.1. One of the great achievements of thermodynamics is the ability to predict the equilibrium state of a system. A system that is out of equilibrium at a given absolute temperature T and pressure p seeks the state (phase and/or chemical composition) that minimizes its total Gibbs free energy,

$$G \equiv H - TS, \tag{1.1}$$

where H is enthalpy and S is entropy. Enthalpy is defined by

$$H \equiv E + pV, \tag{1.2}$$

where E is energy and V is volume.

Thus the driving force for phase change is the departure of G from its minimum value at the given values of pressure, temperature, and total size (volume or mass). The curves separating the phases in Figure 1.1, known as "saturation curves," represent regions of pressure–temperature space in which multiple phases can coexist at equilibrium. When the pressure or temperature changes so as to cross a saturation curve, the system seeks a new equilibrium by undergoing a phase change. However, this process is not instantaneous, because it must be initiated by the formation of nuclei of the new phase, a process that occurs at a finite rate.

Nucleation refers to the initiation of phase change between any two phases. Examples include crystallization from a melt or from an amorphous solid, transition from one crystalline phase to another, bubble formation in a liquid, and condensation of particles from a gas. Traditionally nucleation processes are divided into *homogeneous* nucleation and *heterogeneous* nucleation. Homogeneous nucleation refers to nucleation in the absence of any foreign nuclei or free surfaces. Heterogeneous nucleation occurs when the new phase forms on preexisting foreign nuclei or free surfaces. For example, the formation of solid crystals within the interior of a melt, assuming that the melt is free of void spaces and of preexisting solid particles, is a homogeneous nucleation process, whereas the initial stage

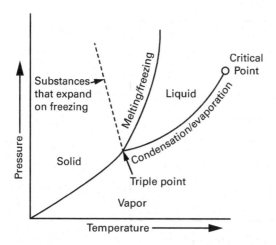

Figure 1.1 Equilibrium phase diagram of a pure substance. For simplicity, the fact that multiple solid phases can coexist for many substances is neglected.

of thin film growth by physical or chemical vapor deposition on a solid surface involves a heterogeneous nucleation process. Both homogeneous and heterogeneous nucleation may be either single-component (homomolecular), involving a single chemical species, or multicomponent (heteromolecular), involving two or more species.

1.2.2   What Is a "Particle"?

Homogeneous nucleation from a vapor leads to the appearance of discrete particles of a condensed phase.

How large must a cluster be before it can be considered a particle? There is no universally agreed definition.

Intuitively, one would want a cluster to be of some minimum size, or to contain some minimum number of atoms or molecules, before it could reasonably be termed a particle.[1] Figure 1.2 shows the relation between the diameter and number n of water molecules in a water cluster, $(H_2O)_n$, assuming that the cluster is spherical and has the same mass density as liquid water at 298 K and 1 atm. These assumptions are obviously simplistic for small values of n, for which an H_2O cluster has a complicated, nonspherical geometry, with an electron cloud whose structure is described by quantum mechanics, and whose volume is thus not well defined. Nevertheless this exercise is instructive.

The calculation is straightforward. An effective molecular volume can be defined by

$$v_1 = \frac{m_1}{\rho} = \frac{\widehat{M}}{\rho N_A}, \tag{1.3}$$

---

[1] Throughout this text, our usage of the term "particle" refers to discrete entities of a condensed phase, an altogether different usage than the elementary "particles" (e.g., atoms, electrons, and other subatomic particles) of physics.

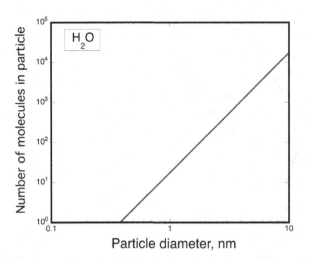

Figure 1.2 Relation between the diameter and number of water molecules in a spherical liquid water cluster.

where m_1 is the mass of one molecule, ρ is the mass density of the bulk condensed phase, $\widehat{M}$ is the molar mass (molecular weight) of the substance, and N_A is the Avogadro constant, 6.022×10^{23} mol^{-1}. Thus v_1 represents the volume occupied per water molecule, in bulk liquid water.

Treating a water molecule as a sphere, the corresponding effective molecular diameter is

$$d_1 = \left(\frac{6v_1}{\pi}\right)^{1/3}. \tag{1.4}$$

For water, we find $d_1 = 0.385$ nm. The number n of molecules contained in a cluster of diameter d_n is then simply

$$n = \left(\frac{d_n}{d_1}\right)^3. \tag{1.5}$$

Based on this analysis, a 1-nm-diameter water "droplet" contains about 17 molecules, but a 4-nm-diameter water droplet contains more than a thousand molecules. At 17 molecules, the assumption of sphericity is dubious, but at a thousand molecules it is probably reasonable, assuming that the droplet is liquid as opposed to an ice crystal.

Let us consider some possible criteria for how large a cluster should be to be called a particle:

(1)   In nucleation theory, the nucleation rate is usually defined as the rate of formation of clusters whose size is just above a so-called "critical size." However, as we shall see, these critical-size clusters may be as small as just a few atoms or molecules or even just a dimer. It seems contrary to common sense to refer to such a small entity as a "particle" of the condensed phase. Indeed, the cluster of

critical size is often termed an "embryo," suggesting that it is not yet a condensed-phase particle but is likely to grow to become one.

(2)   One might identify a cluster as a particle when it is large enough so that some of its properties – for example, density, binding energy per atom, or melting point – are similar to those of the bulk condensed phase of the same substance. However, here one encounters the problem that many properties of nanoparticles, even up to sizes of several tens of nanometers, may be quite different than those of the bulk. For example, for many substances, the melting temperature of particles smaller than about 5 nm in diameter is depressed by several hundred °C compared to the bulk (Goldstein et al. 1992). Many properties of particles smaller than about 10 nm in diameter are strongly affected by the fact that a large fraction of the atoms comprising them lie on or within a monolayer of their surfaces, and for some substances, quantum confinement effects at these sizes can drastically alter their electronic and magnetic properties. Yet it is common to call clusters larger than about one nanometer in diameter "particles," so this test seems inadequate.

(3)   Some researchers involved in computational simulations of chemical nucleation have chosen to define a cluster containing more than an arbitrary number of atoms or molecules – for example, more than 10 or 12 silicon atoms in a silicon hydride (De Bleecker et al. 2004a; Giunta et al. 1990; Swihart and Girshick 1999) – as a "particle," where this definition is made purely for computational convenience.

(4)   It might be argued that a cluster is a "particle" when its structure is that of a solid or a liquid. However, all molecules containing at least two atoms would seem to qualify under this definition, so it makes little sense. A cluster may be "solid-like" or "liquid-like," according to whether it has a well-defined structure and relatively rigid bonds or is instead structureless with its constituent atoms or molecules having considerable mobility. Thus a cluster may experience a melting or freezing transition, and it is meaningful to distinguish between whether it is above or below critical size (and thus whether or not it is stable for growth), but, if it is a constituent of an aerosol, what is meant by saying that it represents a "condensed" phase as opposed to a molecule (even a large one) in the gas phase? The particles in an aerosol, after all, have complete translational freedom, and in most cases, the distances between them are much larger than their individual diameters, similarly as for molecules in a gas.

(5)   An important characteristic of clusters, as noted above, is that a large fraction of their atoms or molecules lie on the surface. One might argue that one should not use the term "particle" until some minimum fraction, for example, 90% or 50%, of the atoms or molecules were located in the "interior" of the cluster as opposed to the surface. To estimate how large the cluster would have to be to satisfy this criterion, suppose for simplicity that the atoms in an atomic cluster were all evenly spaced and arranged in a cube, as in Figure 1.3. Let N denote the width of the cube in number of atoms. The number of atoms in the cluster equals N^3. The fraction f_{int} of atoms located in the interior of the cube is then given by

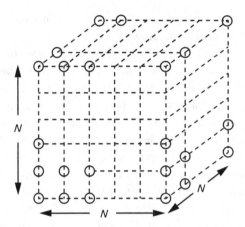

Figure 1.3 Model for the relation between cluster size and the fraction of atoms on the surface. The cube consists of $N \times N \times N$ evenly spaced atoms.

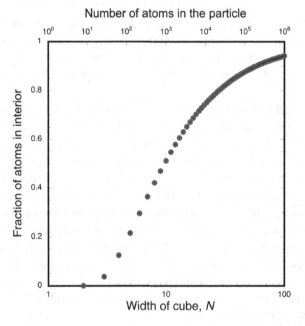

Figure 1.4 Fraction of atoms in the interior (i.e., not on the surface) of a cubic particle vs. number of atoms comprising the width of the cube.

$$f_{\text{int}} = \frac{(N - 2)^3}{N^3}. \tag{1.6}$$

Figure 1.4 graphs the result. One finds that the 50% criterion is not satisfied until $N = 10$, corresponding to a cluster of 1,000 atoms, and the 90% criterion is not satisfied until $N = 58$, corresponding to a cluster of almost 2×10^5 atoms. The corresponding width of the cube in nanometers would depend on the size of the atoms and the spacing between them. If one assumed, for example, that each atom, accounting for the spacing, occupied a lateral dimension of 0.3 nm, then over 50% of the atoms would lie on the surface for cube widths smaller than 3.0 nm, and over 10% of the atoms would lie on the surface for cube widths smaller than 17.4 nm. As most current usage would accept the term "particle" for sizes somewhat smaller than 3 nm, it is evident that one must recognize that "particle" is used to denote entities for which a very large fraction of atoms lie on the surface and which therefore may have properties that are quite different from those of the bulk condensed phase. Indeed, this is a major reason for the strong interest in nanoparticles for a variety of technological applications.

(6) Taking account of the problems with (4) and (5), one could reserve the term "particle" for a cluster for which the structure of the interior atoms was identical to that of a bulk crystalline phase of that substance. It does seem reasonable to call such a cluster a "particle" and even to acknowledge that in this case a phase transition appears to have occurred. However, this definition would be too restrictive, as one also wants to use the term "particle" to describe liquid droplets and amorphous solids.

(7) Experimentalists studying nucleation rates may for practical reasons define the minimum size of a "particle" as the smallest cluster in an aerosol that can be detected by methods such as laser light scattering or instruments such as differential mobility analyzers. In this case, the minimum size of a "particle" depends on the detection limit of the method and/or instrument. For example, at present the lower detection limit for differential mobility analyzers equals about 1 nm, suggesting a practical experimental demarcation between "clusters" and "particles."

(8) Given the above, a reasonable choice for the minimum diameter of a "particle" might be taken as 1 nm.[2] This choice emphasizes the approximate nature of the division between clusters and particles and is clearly the order of magnitude in the SI unit system that best corresponds to the lower limit of particle size. Nevertheless, one must keep in mind that a 1-nm "particle" contains very few atoms or molecules and may be highly nonspherical, with an ill-defined volume. Indeed, in many cases large organic molecules are themselves larger than 1 nm.

In summary, what constitutes a particle of a condensed phase is not well defined. However, this does not present a problem for gas-phase nucleation theory, because it focuses not on the formation of "particles" but on the formation of nuclei – the entities that are likely to grow to the size of particles. And, as shall be seen, "nuclei" *are* rigorously defined.

For a further discussion of length scales that are pertinent to the transition from "molecules" to "particles," the reader is referred to an interesting paper by Preining (1998).

---

[2] We use the term "diameter" in the generic sense to denote a characteristic linear dimension of a particle that is not necessarily spherical.

1.2.3   Single-Component Homogeneous Nucleation from the Gas Phase

Phase-change processes between any pair of phases have enough in common to be considered within a general theory of nucleation. A comprehensive text on the general theory of nucleation is *Nucleation*, by Kashchiev (2000). However, each type of nucleation also presents a number of issues that are unique to the phases involved. In this text, we are concerned with the nucleation of condensed-phase particles from gases. An earlier textbook focused on this subject is *Homogeneous Nucleation Theory: The Pretransition Theory of Vapor Condensation*, by Abraham (1974).

The simplest case involves homogeneous nucleation of a single-component supersaturated vapor, which can occur when the saturation curve in Figure 1.1 is crossed from the region marked "vapor" to either of the regions marked "solid" or "liquid."[3] This process is often termed "self-nucleation," meaning that the vapor condenses on itself, and the term "homomolecular" is sometimes used to indicate that only one molecular species condenses.

Consider, for example, water vapor at equilibrium at a pressure of 1 atm and a temperature of 101°C that is cooled at constant pressure to a temperature of 99°C, as illustrated in Figure 1.5. By definition of the Celsius scale, the equilibrium vapor pressure of water equals 1 atm (101.325 kPa) at 100.0°C. At 101°C, the equilibrium vapor pressure of water equals 105.0 kPa; at 99°C, it equals 97.76 kPa (Haar et al. 1984).

The saturation ratio, S, of a vapor is defined by

$$S = \frac{p}{p_s(T)}, \tag{1.7}$$

where p is the actual partial pressure of the vapor and p_s is the equilibrium (or "saturation") vapor pressure.[4]

At the state marked "1" in Figure 1.5, $S = 101.3/105.0 = 0.965$; at State 2, $S = 1$; and at State 3, $S = 101.3/97.76 = 1.04$. If the system at each of these states were given sufficient time to reach equilibrium, State 1 would be vapor, State 2 would be a mixture of vapor and liquid – the proportion of each depending on the specific volume – and State 3 would be all liquid. However, if this experiment were conducted in such a way that effectively no walls or preexisting particles existed, one would find that one could substantially overshoot the saturation curve before any condensation was observed. Upon crossing the saturation curve from the vapor side to the liquid side, S becomes greater than unity, and the vapor is said to be "supersaturated." In this case, the system is out of equilibrium, and thermodynamics tells us that a driving force for phase change exists. However, nucleation is a kinetic process, and the time required for observable condensation depends on various conditions. In many cases, it is possible for vapors to

---

[3] While the words "vapor" and "gas" can be used interchangeably, the use of "vapor" emphasizes that the gas in question is condensible under the conditions of interest.

[4] The literature is inconsistent on terminology here. We will usually use "saturation ratio," as in standard textbooks on aerosol science (Friedlander 2000; Hinds 1999; Seinfeld and Pandis 1998), although one often finds S referred to as the "supersaturation," which strictly speaking should refer to the quantity $(S - 1)$. One also finds the terms "supersaturation ratio" and the "degree of supersaturation."

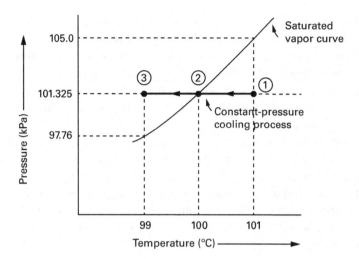

Figure 1.5 Constant pressure cooling of water, causing it to cross the liquid–vapor saturation curve.

be highly supersaturated (e.g., saturation ratios of several hundred or greater) without observable condensation over laboratory time scales. C. T. R. Wilson, who conducted cloud chamber experiments for which he won the 1927 Nobel Prize in Physics, observed no nucleation in water vapor free of ions at room temperature until the saturation ratio reached a value of about eight (Wilson 1897). Indeed, the classical theory of single-component homogeneous nucleation, discussed in Chapter 3, predicts for water vapor at 20°C and at a relative humidity of 200% (i.e., $S = 2$) that the steady-state nucleation rate equals about 10^{-54} cm^{-3} s^{-1}, which works out to equal about one water droplet per cm^3 per 10^{47} years. The estimated life of the universe being only about 10^{10} years, one is not likely to have observed that droplet.[5]

The physical nature of the self-nucleation process is that molecules of the vapor collide with each other and stick together to form larger and larger clusters. In general, these clustering collisions are reversible, that is, a molecule can "evaporate" from a cluster and return to the vapor phase. In fact, until a cluster grows to a certain critical size, the value of which depends on the properties of the substance and on conditions, it is more likely to shrink than to grow. But should it succeed in growing past the critical size, further growth becomes effectively irreversible. The existence of a critical size larger than a monomer thus poses a barrier to nucleation, and clusters of critical size represent the "nuclei" for phase change.[6]

---

[5] The steady-state assumption would obviously not be valid here, and even if it were, the classical theory can be in error by many orders of magnitude, but even so this calculation gives a qualitative sense of how rare homogeneous nucleation of water would be under these conditions.

[6] In organic chemistry, the term "monomer" is used to mean a simple organic molecule that can join in long chains with other molecules to make a polymer. Here we use the term "monomer" more generally to mean the smallest entity, not necessarily organic, having the same chemical composition as the nucleating substance. In most cases, a "monomer" is simply a single molecule (or atom, if the molecule is monatomic) of the condensible vapor.

Nucleation is the process of forming these nuclei. Note that the nuclei themselves are not necessarily entities of the new phase. They are simply molecular or atomic clusters, often just large (or even not so large) molecules. Per the preceding discussion, entities large enough to be considered particles of the condensed phase may be much larger than the nuclei. However, as the nuclei are by definition the entities that are stable for growth, the rate of particle formation can be closely related to the rate of nuclei formation – the nucleation rate.

According to the value of the critical size, one may distinguish between two different regimes. Most of the literature on the theory of homogeneous nucleation is concerned with the scenario discussed above, in which the critical size is larger than a monomer and growth to the critical size requires passage through a sequence of reversible clustering reactions, each of which has a higher rate in the direction of decay than in the direction of growth. This scenario describes what is known as the "condensation/evaporation" regime.

However, in cases of extremely high supersaturation, the critical size may be as small as the monomer itself. This could occur, for example, when a substance with very low equilibrium vapor pressure is generated at high rates by chemical reactions. In that case, the saturation ratio could shoot to an extremely high value in a time shorter than is required for significant clustering to occur. For example, powders of ceramic materials can be synthesized by generating the monomer vapor in a flame or plasma. In some cases, the chemical generation of the vapor occurs at a temperature where the equilibrium vapor pressure is extremely low. In this case, there might effectively be no barrier to nucleation, and the nucleation rate is limited only by the rate of monomer–monomer collisions that form dimers. This situation is sometimes termed the "collision-controlled regime" (McMurry and Friedlander 1979; Rao and McMurry 1989) or "collisional limit" (Dingilian et al. 2021).

1.2.4   Other Types of Nucleation from the Gas Phase

As noted above, in general the driving force for phase change is the departure of the total free energy of a system from its minimum value at given temperature and pressure. Homomolecular homogeneous nucleation is a special case of this, in which the clustering involves a single chemical species.[7]

It is also observed that multiple vapors can nucleate together, with the nuclei being composed of a mixture of two or more substances, and that this can occur under conditions where neither substance would nucleate by itself. In general, this is termed "heteromolecular" or "multicomponent" nucleation, with specific scenarios being denoted "binary nucleation," "ternary nucleation," etc., depending on how many species of condensing vapors are involved. Examples that are important for atmospheric aerosols include the binary nucleation of sulfuric acid and water and the ternary nucleation of sulfuric acid, ammonia, and water.

---

[7] We will always use "free energy" to mean "Gibbs free energy," defined by equation (1.1), not to be confused with the Helmholtz free energy, $F \equiv E - TS$, where E is internal energy.

The presence of ions can lower the degree of vapor supersaturation required for observable nucleation to occur. Thus ions are said to "induce nucleation." Ion-induced nucleation is conventionally classified in the literature as a heterogeneous nucleation process, in the sense that ions are viewed as constituting preexisting seeds onto which a vapor condenses. Note, however, that this usage of the term "heterogeneous" is peculiar to nucleation, as ion-molecule reactions occurring in the gas phase are considered homogeneous processes from the viewpoint of chemical kinetics. Indeed, ion-induced nucleation is a special case of gas-phase nucleation and thus fits within this book on the nucleation of particles from the gas phase.

On the other hand, in the growth of thin films by physical or chemical vapor deposition on solid surfaces, the initial stage of film growth involves the nucleation of small islands on the surface, which ultimately may grow to comprise a continuous film. Nucleation here is inherently a heterogeneous process, involving reactions of gas molecules on a solid surface. A similar situation pertains to nucleation that occurs as the initial phase of vapor condensation on preexisting condensed-phase particulates, for example, raindrop formation on cloud condensation nuclei. Thus these processes would not be considered "gas-phase nucleation" and do not fall within the scope of this book.

Most of the literature on homogeneous nucleation assumes implicitly that nucleation is a "physical" condensation process, that is, involving physisorption not chemisorption, and that the molecules comprising the critical nucleus are held together by relatively weak van der Waals forces, as opposed to chemical bonds. However, the increasing appreciation and understanding of chemically bound clusters, and of systems in which clusters grow by chemical reactions, has made "chemical nucleation" a fertile area of research. An important example is the nucleation of soot during hydrocarbon combustion. It should be noted that the term "chemical nucleation" has sometimes been applied to processes in which a supersaturated vapor is generated by chemical reactions and then nucleates via cluster growth that does not itself involve chemical reactions (e.g., Katz and Donohue 1982). Indeed, if the rate of chemical generation of the vapor is high enough, then the subsequent nucleation may be in the collision-controlled regime described above, in which case the rate-limiting process may be the rate of vapor generation by chemical reactions itself. In this text, we reserve the term "chemical nucleation" for processes in which clusters are chemically bound entities that grow by chemical reactions.

Another type of gas-phase nucleation discussed in this text is nucleation in plasmas, which can involve aspects of both ion-induced nucleation and chemical nucleation, as well as a wealth of phenomena that are not typically considered in nucleation theory. Plasmas that contain condensed-phase particles, termed "dusty plasmas," are the subject of a considerable body of literature. Much of the universe is in a plasma state, and nucleation of dust in space under plasma conditions may play a key role in the birth of stars and other astrophysical bodies. Plasmas are widely used in semiconductor processing for microelectronics fabrication. As the characteristic feature sizes of integrated circuits shrink to nanometer dimensions, avoidance of particle nucleation in processing plasmas presents an important technological challenge, because deposition of particles on processing wafers can cause unacceptable defects. On the other hand, plasmas are increasingly used for the deliberate synthesis of nanoparticles for various applications.

1.3   Scope and Level of This Text

This text is concerned with the transition of a system that is completely in the gas phase to one that includes condensed-phase particles, dispersed in the gas phase – an "aerosol." Thus, although much of the material presented here may be more broadly applicable, this text does not consider other phase transitions, such as crystallization of a melt, bubble formation in liquids, and so forth.

This text adopts the viewpoint that nucleation can and should be considered within the general framework of thermochemistry and chemical kinetics, regardless of whether cluster growth actually involves the making and breaking of chemical bonds. Classical theories of nucleation, which model clusters as having the same properties as the bulk condensed phase, are often presented using rather different concepts and terminology than would be natural from the viewpoint of thermochemistry and kinetics. However, classical theories are seen to fit perfectly well within the broader context of thermochemistry and kinetics and indeed are based on reasonable approximations for those cases where necessary data on cluster properties are unavailable. At the time of this writing, that is still most often the case, and classical models are still widely used for estimating nucleation rates for many kinds of systems. Therefore, both for completeness and because they are still in widespread use, this text presents detailed discussions of the classical theories of single-component, multicomponent, and ion-induced nucleation.

One can confidently predict, however, that atomistic approaches will increasingly be used, as more data become available on the properties and kinetics of clusters. This text does not attempt to describe in detail the methods of computational or experimental chemistry that are used to generate these data. It does, however, provide a framework for how these data can be used within models for various types of particle nucleation from the gas phase, and considers ways in which atomistic approaches might predict qualitative differences in our picture and understanding of nucleation.

We hope that this text will be of interest to specialists in the field of nucleation, but we intend for it to be comprehensible to senior undergraduates and beginning graduate students in physical sciences and engineering, and to researchers in broader fields such as aerosol science and materials synthesis for whom nucleation is an important topic. Thus we do not assume any prior knowledge of nucleation, only that the reader has a good basic undergraduate education in some area of the physical sciences or engineering.

Homework Problems

In these and other homework problems throughout the text, it is assumed that you can find any necessary property data in suitable reference sources, including widely accessible online sources such as Wikipedia. Always state the source of your data.

1.1   Based on the model presented in Section 1.2.2, what is the diameter of one molecule of ethanol at 20 °C?

1.2   Figure 1.2 plots the number of water molecules in a spherical liquid water cluster or nanoparticle as a function of particle diameter. Prepare a similar graph for silicon, assuming the same mass density as for bulk silicon at room temperature, 2.33 g·cm^{-3}. Based on your calculations, what is the number of silicon atoms in spherical silicon particles having diameters of 1, 10, and 100 nm?

1.3 Assuming the same mass density for silicon as in the previous example, consider a silicon cube measuring 2 nm on a side. What fraction of the Si atoms are located on the surface? Note that 2 nm may not be an exact multiple of the interatomic spacing, so make a suitable approximation.

1.4 Consider water vapor at a pressure of 10 kPa. How does the saturation ratio change as the temperature changes from 50 °C to 10 °C, keeping the pressure constant?

2 Single-Component Homogeneous Nucleation from a Supersaturated Vapor

We first consider the simplest problem in gas-phase nucleation: homogeneous nucleation of a single substance from a supersaturated vapor. The theoretical treatment of this problem has evolved and has been presented in the literature in various forms. Seemingly disparate derivations have been followed by various authors to produce essentially identical results, a situation that has bred a great deal of confusion. We here present the theory in a way that does not attempt to replicate its historical evolution or even the standard presentations, but which rather, we believe, is the most natural and straightforward, and most amenable, subsequently, to replacing classical assumptions with atomistic approaches. Thus we first derive an expression for the steady-state nucleation rate in a general form that is independent of the model used for cluster properties.

2.1 Problem Description

We consider a system that is initially entirely in the gas phase, consisting of a supersaturated vapor of a single substance. The vapor may be one component in a mixture of gases, but if so, only that one vapor species is supersaturated, none of the other species is close to being condensible under the pressure and temperature conditions involved, and all other species are inert as far as chemical reactions with the condensible vapor are concerned. There are no preexisting stable nuclei and no walls or other surfaces.

Such a system is obviously an idealization. For example, air in urban environments typically contains 10^4–10^5 aerosol particles per cm^3 (Friedlander 2000). But in fact the idealization can come close to being realized in many actual situations outside contrived laboratory experiments. Examples include regions of the marine atmosphere that are nearly free of particles, supersonic nozzle expansions where the timescales in the flow direction during nucleation are much shorter than for radial transport to walls, the interior regions of flames and plasmas, and so forth. Even if there are preexisting particles, homogeneous nucleation can still occur if the concentration of those particles is sufficiently low (Pesthy et al. 1981).

If the vapor is supersaturated, then by definition the system is out of equilibrium. Unless external inputs exist to maintain this nonequilibrium, the system must therefore change until equilibrium is achieved by the vapor condensing until its supersaturation is relieved. This condensation process is initiated by the vapor molecules clustering to form stable nuclei. Our primary goal is to develop a theory for predicting the rate of formation of these nuclei.

If one assumes that the supersaturated vapor exists initially as a collection of dispersed molecules, then the formation of clusters must involve collisions among these molecules in which they stick together. We will refer to such a sticking collision as a "reaction," regardless of whether the sticking actually involves the formation and/or breaking of chemical bonds or instead is just a physisorption process, and will assume that it constitutes an elementary reaction.

One can hypothesize that the growth of clusters occurs by a sequence of such reactions. Let A_n denote a cluster of species A that contains n monomers, or an "n-mer." We adopt the viewpoint that all of the A_n's ($n = 1, 2, 3, \ldots$) can themselves be treated as distinct chemical species, again regardless of whether they represent chemically bound or "physically" bound entities. In other words, clusters in this formalism are treated no differently than molecules. Then the sequence of reactions to form A_n might be written as

$$A_1 + A_1 \rightarrow A_2 \tag{2.1}$$

$$A_2 + A_1 \rightarrow A_3 \tag{2.2}$$

$$A_3 + A_1 \rightarrow A_4 \tag{2.3}$$

$$\vdots$$

$$A_{n-1} + A_1 \rightarrow A_n. \tag{2.4}$$

One could also have reactions that do not involve monomers, for example,

$$A_2 + A_3 \rightarrow A_5. \tag{2.5}$$

However, under many conditions it is reasonable to assume that monomers are far more abundant than clusters of any other size, at least until monomers have become substantially depleted, which usually does not happen until a significant amount of particle nucleation and growth has already occurred. Under this assumption, a cluster of any size is far more likely to collide with a monomer than with a dimer or any larger cluster. Therefore we will neglect, unless stated otherwise, all reactions except those involving a monomer as one of the collision partners.

Even though our system is out of equilibrium, with a superabundance of monomers relative to clusters of all larger sizes, one has no a priori reason to think that any of the reactions in this sequence is irreversible. So the sequence should properly be rewritten as follows:

$$A_1 + A_1 \rightleftarrows A_2 \tag{2.6}$$

$$A_2 + A_1 \rightleftarrows A_3 \tag{2.7}$$

$$A_3 + A_1 \rightleftarrows A_4 \tag{2.8}$$

$$\vdots$$

$$A_{n-1} + A_1 \rightleftarrows A_n. \tag{2.9}$$

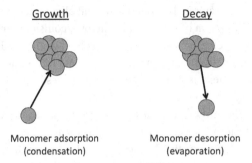

Growth

Decay

Monomer adsorption
(condensation)

Monomer desorption
(evaporation)

Figure 2.1 Cluster growth and decay by monomer adsorption/desorption.

From the viewpoint of chemical kinetics, if the reactions shown are elementary reactions, then the reactions in the forward direction are bimolecular association reactions, and the reactions in the reverse direction are unimolecular dissociation reactions. Strictly speaking, however, these are not elementary reactions, and one should explicitly show the collision partner required in the forward direction to stabilize the excited complex, or transition state, formed in the collision, and in the backward direction to cause dissociation. For example, reaction (2.6) could instead be written as

$$A_1 + A_1 + M \rightleftarrows A_2 + M, \tag{2.10}$$

where M stands for any collision partner. Nucleation theory overwhelmingly neglects this detail, which is equivalent to assuming that all reactions of the form of (2.9) are in their high-pressure limit, in which collision partners are sufficiently abundant that the rate constants for the forward and backward directions do not depend on the total pressure of the system. We return to this point in Chapter 6, but for now we will continue to write these reactions as in (2.9).

As a corollary of the assumption that cluster growth occurs by monomer addition only, it can be noted that cluster decay is assumed to occur by monomer evaporation only.

The situation is illustrated in Figure 2.1. Clusters grow by addition (condensation) of single monomers and shrink by loss (evaporation) of single monomers.

2.2   Reaction Rates

2.2.1   Forward and Backward Rate Constants

Reaction (2.9) is the general form of the reversible reaction for condensation/evaporation. Let R_n denote the forward rate of this reaction, and let R_{-n} denote the reverse rate. The units of R_n and R_{-n} are defined such that the net rate of change in the n-mer number density N_n (n-mers per unit volume) due to reaction (2.9) is given by

$$\frac{dN_n}{dt} = R_n - R_{-n}, \quad n \geq 2. \tag{2.11}$$

It should be emphasized that this equation gives the rate of change of N_n due solely to reaction (2.9) and that in general there may be other reactions and local sources and/or sinks that contribute to dN_n/dt. We return to this point in Section 2.4 and Chapter 7.

As we assume that reaction (2.9) is an elementary reaction in both directions, the reaction rates are proportional to the number densities of each of the reacting species, that is, the reactions are first order with respect to each of these species. The proportionality constant is called a "rate constant" (although in general it is not constant) or "rate coefficient." Thus the forward and backward rates of reaction (2.9) can be rewritten as

$$R_n = k_n N_{n-1} N_1, \ n \geq 2 \tag{2.12}$$

and

$$R_{-n} = k_{-n} N_n, \ n \geq 2, \tag{2.13}$$

where k_n and k_{-n} are the rate constants for the forward and backward (or reverse) directions, respectively. As reaction (2.9) is not equimolar, k_n and k_{-n} have different units. The units of k_n are $cm^3 \cdot s^{-1}$, while the units of k_{-n} are s^{-1}. As the forward and reverse reactions represent monomer addition and monomer removal from the n-mer, they are often referred to as "condensation" and "evaporation," respectively, regardless of the fact that whether or not the n-mer represents a small piece of the condensed phase is not well defined.

The rate constants in general depend on temperature, and in some cases on pressure as well, as discussed in Chapter 6.

The rate constants for the forward and backward directions are not independent of each other. For the special case where reaction (2.9) is in equilibrium, the net reaction rate must by definition equal zero. Equating the right-hand sides of equations (2.12) and (2.13), it follows that

$$\frac{k_n}{k_{-n}} = \left(\frac{N_n}{N_{n-1} N_1} \right)_{eq}, \ n \geq 2, \tag{2.14}$$

where the subscript "eq" denotes that the number densities of the species involved are equilibrated with each other at the given temperature and pressure.

2.2.2   The Equilibrium Constant for Condensation/Evaporation

The right-hand side of equation (2.14) is recognized as the equilibrium constant for reaction (2.9), in number density form. To express this in terms of a dimensionless equilibrium constant K, let us assume that the monomers and all n-mers behave like separate species in a mixture of ideal gases, so that each obeys the ideal gas law,

$$p_n = N_n k_B T, \tag{2.15}$$

where p_n is the partial pressure of n-mers and k_B is the Boltzmann constant, 1.381×10^{-23} J·K^{-1}. Using this, one can write

$$\left(\frac{N_n}{N_{n-1}N_1}\right)_{eq} = k_B T \left(\frac{p_n}{p_{n-1}p_1}\right)_{eq}, \ n \geq 2. \tag{2.16}$$

The dimensionless equilibrium constant K for reaction (2.9) is defined by normalizing each of the partial pressures by an arbitrary reference or "standard state" pressure, p^0, conventionally chosen as 1 atm or 1 bar:

$$K(T) \equiv \left[\frac{p_n/p^0}{(p_{n-1}/p^0)(p_1/p^0)}\right]_{eq}, \ n \geq 2. \tag{2.17}$$

If the species participating in the reaction behave as ideal gases, then from thermodynamics K is a function only of temperature. Unless stated otherwise, we will use this dimensionless form of the equilibrium constant. To emphasize that the equilibrium constant defined by equation (2.17) pertains specifically to reaction (2.9), we will write it as $K_{n-1,n}$, meaning the equilibrium constant for the reaction in which a cluster grows from size $(n-1)$ to size n by the addition of a monomer.

Comparing equations (2.16) and (2.17), one thus has

$$\left(\frac{N_n}{N_{n-1}N_1}\right)_{eq} = \left(\frac{k_B T}{p^0}\right) K_{n-1,n}, \ n \geq 2. \tag{2.18}$$

From equation (2.15), let us define a "standard number density" N^0 for an ideal gas at temperature T by

$$N^0(T) \equiv \frac{p^0}{k_B T}. \tag{2.19}$$

Then, comparing equations (2.14) and (2.18), one can write

$$\frac{k_n}{k_{-n}} = \frac{K_{n-1,n}}{N^0}, n \geq 2. \tag{2.20}$$

2.2.3   The Stepwise Change in Standard Gibbs Free Energy

From thermodynamics, one further has the general result, for ideal gases, that the equilibrium constant can be expressed in terms of ΔG^0, the change in Gibbs free energy for a reaction in which the free energies of all reactant and product species are evaluated at standard pressure:

$$K(T) = \exp\left[-\frac{\Delta G^0(T)}{k_B T}\right], \tag{2.21}$$

where ΔG^0 here is in molecular units.[1] Thus the equilibrium constant for the condensation/evaporation reaction, (2.9), can be written as

---

[1] In this text, quantities such as G are on a "per molecule" basis unless stated otherwise. If using molar units, replace the Boltzmann constant in equation (2.21) and similar expressions with the molar gas constant, $R = N_A k_B$.

$$K_{n-1,n} = \exp\left(-\frac{\Delta G^0_{n-1,n}}{k_B T}\right), \tag{2.22}$$

where $\Delta G^0_{n-1,n}$ is known as the "stepwise" change in standard Gibbs free energy, as it pertains to the reaction in which a cluster grows by the addition of a single monomer.

Equation (2.20) for the ratio of the rate constants was derived by assuming that the reaction to which it pertains is in equilibrium. However, a common and usually valid assumption of chemical kinetics is that, for a given temperature, the values of the rate constants for elementary reactions do not depend on whether or not the system is in equilibrium (Benson 1976; Vincenti and Kruger 1965). Therefore, it is evident from equations (2.20) and (2.22) that if one knows either of the rate constants k_n and k_{-n}, then the other can be found if one knows the stepwise standard free energy change for the reaction that forms an n-mer.

Usually in nucleation theory, it is the rate constant for condensation that is known. In that case, the rate constant for evaporation from an n-mer can be written as

$$k_{-n} = k_n N^0 \exp\left(\frac{\Delta G^0_{n-1,n}}{k_B T}\right), \ n \geq 2. \tag{2.23}$$

2.3   The Gibbs Free Energy of Cluster Formation from the Monomer Vapor

To evaluate changes in Gibbs free energy for reactions that involve condensation/ evaporation of monomers onto/from n-mers, it is useful to define a "Gibbs free energy of cluster formation from the monomer vapor" for n-mers of arbitrary size. By convention, the Gibbs free energy of formation, $\Delta_f G$, of any chemical substance is defined as the change in free energy associated with forming the substance from its constituent "standard reference elements," for a hypothetical reaction in which each of the reactants and the product of the reaction (the substance itself) are at a common temperature and pressure. The "standard reference elements" are defined as the chemical elements that comprise the substance, in their most stable form at pressure p^0 and at the temperature at which $\Delta_f G$ is evaluated. For example, to define the free energy of formation of gaseous methane at 25°C, one evaluates the free energy change associated with the reaction

$$C(s) + 2H_2(g) \rightarrow CH_4(g), \tag{2.24}$$

where "s" and "g" denote solid and gas phases, respectively, as solid carbon – specifically, the graphite phase – and gaseous H_2 are the most stable forms of carbon and hydrogen, respectively, at 25°C and 1 bar.

For the problem at hand, however, it is more useful to define a free energy of cluster formation somewhat differently, because by definition single-component homogeneous nucleation involves the formation of clusters from a monomer vapor of the same chemical substance. Therefore, the free energy of formation we wish to assign to an

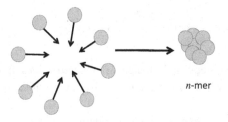

n monomers in vapor

Figure 2.2 Overall reaction used to define Gibbs free energy of cluster formation.

n-mer is the change ΔG associated with forming an n-mer from n monomers of the same chemical substance,

$$n\mathrm{A}_1 \rightarrow \mathrm{A}_n, \tag{2.25}$$

where the monomers are by definition in the vapor phase, regardless of the pressure and temperature. This reaction, illustrated in Figure 2.2, is simply the overall reaction expressed by the sequence of condensation reactions, reactions (2.6)–(2.9). We will denote the free energy change associated with reaction (2.25) by ΔG_n and will refer to it as the "free energy of cluster formation from the monomer vapor," or simply the "free energy of cluster formation." It should not be confused with the free energy of formation of a chemical substance, $\Delta_f G$.

The free energy of cluster formation from the monomer vapor is simply equal to the sum of all the stepwise free energy changes involved in growing a cluster from the size of a monomer to the size of an n-mer:

$$\Delta G_n = \sum_{i=2}^{n} \Delta G_{i-1,i}, \; n \geq 2. \tag{2.26}$$

Alternatively, following our convention for the definition of ΔG_n, we can write it as

$$\Delta G_n(T,p) = G_n(T,p) - n\tilde{G}_v(T,p), \; n \geq 1, \tag{2.27}$$

where

$$\tilde{G}_v = G_1 \tag{2.28}$$

denotes the free energy per molecule in the vapor phase. If evaluated at standard pressure, the free energy of cluster formation from the monomer vapor is written as $\Delta G_n^0(T)$.

As an aside, it should be noted that equation (2.27) automatically sets the free energy of formation from the monomer vapor of a monomer itself to zero. It may seem obvious that this should be so, as, from equation (2.25), ΔG_1 represents the free energy change for the trivial reaction

$$\mathrm{A}_1 \rightarrow \mathrm{A}_1. \tag{2.29}$$

However, much of the nucleation literature has historically adopted models in which $\Delta G_1 \neq 0$, which leads to various "self-consistency" problems. This issue is discussed further in Section 3.10. We take the viewpoint that *whatever* the model used to evaluate ΔG_n, it must at least be consistent with the identity

$$\Delta G_1 \equiv 0. \tag{2.30}$$

2.4   The Nucleation Current

We return now to consideration of the clustering sequence, reactions (2.6)–(2.9). Let J_n denote the "nucleation current," defined as the net rate per unit volume of n-mer formation associated specifically with reaction (2.9). The nucleation currents at each size starting with the dimer can be written as

$$J_2 = k_2 N_1 N_1 - k_{-2} N_2 \tag{2.31}$$

$$J_3 = k_3 N_2 N_1 - k_{-3} N_3 \tag{2.32}$$

$$J_4 = k_4 N_3 N_1 - k_{-4} N_4 \tag{2.33}$$

$$\vdots$$

$$J_n = k_n N_{n-1} N_1 - k_{-n} N_n = R_n - R_{-n}. \tag{2.34}$$

As illustrated in Figure 2.3, the nucleation current J_n represents the net transfer rate out of size $(n-1)$ and into size n, while J_{n+1} represents the net transfer rate out of size n and into size $(n+1)$, due solely to condensation/evaporation reactions.[2] Thus the net rate of change of the n-mer number density due to all condensation/evaporation reactions is given by

$$\frac{dN_n}{dt} = J_n - J_{n+1}, \; n \geq 2. \tag{2.35}$$

This equation represents the total rate of change in the n-mer number density for a closed system (no n-mer sources or sinks due to transport of n-mers across system boundaries), assuming that the number of cluster–cluster reactions is negligible compared to that of condensation/evaporation events.

The use of the term "current," in analogy with flow of water in a river or of electrical charge in a conductor, thus represents a net rate of transfer from one location (in this case, in cluster size space) to an adjacent location, without necessarily implying any net accumulation (in n-mer number density, water level, or electrical charge) at that location, unless the currents at adjacent locations differ.

From equation (2.23), the backward rate constants that appear in equations (2.31)–(2.34) can be eliminated in favor of the forward rate constants and equilibrium constants for each reaction, giving

---

[2] To emphasize this point, the nucleation current is sometimes written as $J_{n-1/2}$ and $J_{n+1/2}$. For our purposes, this notation is unduly cumbersome, but the reader should keep in mind that J_n represents a net transfer rate "from below."

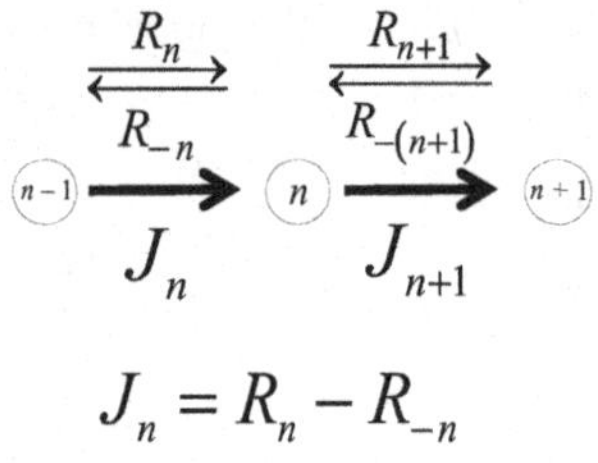

$$J_n = R_n - R_{-n}$$

Figure 2.3 Schematic illustration of nucleation current.

$$J_2 = k_2 N_1 N_1 - \frac{k_2}{K_{1,2}/N^0} N_2 \tag{2.36}$$

$$J_3 = k_3 N_2 N_1 - \frac{k_3}{K_{2,3}/N^0} N_3 \tag{2.37}$$

$$J_4 = k_4 N_3 N_1 - \frac{k_4}{K_{3,4}/N^0} N_4 \tag{2.38}$$

$$\vdots$$

$$J_n = k_n N_{n-1} N_1 - \frac{k_n}{K_{n-1,n}/N^0} N_n \tag{2.39}$$

2.5 The Steady-State Approximation

Equation (2.35) expresses the rate of change of n-mer number densities as the difference between the nucleation currents for adjacent sizes. In general, the nucleation currents (2.36)–(2.39) can vary in time. Thus

$$\frac{dN_n}{dt} = J_n(t) - J_{n+1}(t), \ n \geq 2. \tag{2.40}$$

However, in many cases of practical interest, it is reasonable to assume that an "instantaneous steady state" prevails such that

$$\frac{dN_n}{dt} = 0 \tag{2.41}$$

for clusters of all sizes greater than the monomer. In that case, it follows from equation (2.40) that

$$J_2(t) = J_3(t) = J_4(t) = \cdots = J_n(t) = \cdots = J_M(t) = J(t), \tag{2.42}$$

where M represents some arbitrarily large size up to which the steady-state nucleation current can be assumed to hold.

By "steady state," we mean that the cluster size distribution N_n adjusts extremely rapidly to changes in the forcing conditions (such as temperature and vapor saturation ratio), that is, it adjusts on timescales that are very short compared to the timescales for the changes in those conditions. Therefore, even though changes in forcing conditions can cause the nucleation currents to change with time, the cluster size distribution rapidly adjusts to any new set of forcing conditions to establish a new "steady state." This situation is analogous to the "pseudo-steady-state" approximation applied to specific reactions in chemical kinetics, in which a given reaction or set of reactions can be assumed to be in steady state, because the reactions involved are much faster than other reactions, changes of state, or transport terms that affect the concentrations of the species involved.

The analogy to river flow is again instructive. Consider a river fed by a large reservoir. Over the course of a year, the river flow rate and the water level may change considerably, but on any given day, the level at any given location may be effectively constant in time, and, by mass continuity, assuming no external sources or sinks, the flow rates across every cross section of the river normal to the flow direction must equal each other.

For many systems, the time required to reach steady-state nucleation is believed to be quite short compared to timescales for system change. For now, we assume that steady state prevails. This results in a great simplification of the nucleation problem. Situations where steady-state nucleation is not a valid assumption, and the general approach to transient nucleation, are discussed in Chapter 7.

2.6   The Steady-State Nucleation Rate in Summation Form

Assuming the existence of steady state as applied to equations (2.36)–(2.39), the subscripts can be dropped from the nucleation currents. Therefore J here represents the steady-state (or "stationary") nucleation rate.

We can then write a sequence of equations for the steady-state nucleation rate, all of which have the form

$$J = k_n N_{n-1} N_1 - \frac{k_n}{K_{n-1,n}/N^0} N_n, \ 2 \le n \le M. \tag{2.43}$$

As we have $(M - 1)$ such equations, where M is arbitrarily large, we have an arbitrarily large number of equations for the quantity J whose value we are trying to find, as well as for the unknown cluster number densities. Solving this problem, it would thus appear, involves solving not only for J but also for the cluster size distribution N_n that exists, from equation (2.41), during steady-state nucleation, and that is indeed an approach often found in the literature. In fact, however, while the cluster size distribution that corresponds to steady-state nucleation may be of interest in its own right, it is not necessary for the problem at hand – finding the value of J – because the sequence of equations can be reduced to a single equation in which all of the N_n for sizes larger than a monomer and smaller than an M-mer are eliminated.

Note that steady state is not the same as equilibrium; the steady-state cluster size distribution is not the same as the equilibrium cluster size distribution. If the cluster size distribution were in equilibrium, the net forward rate for each reaction in the clustering sequence would equal zero, that is, all the nucleation currents would equal zero and no nucleation would occur. Nucleation is driven by a nonequilibrium cluster size distribution.

To solve for the steady-state nucleation current, we seek an appropriate recursion relation to reduce all the equations in the sequence (2.43) to a single equation. To begin, let us divide both sides of every equation by the factor $k_n N_1$, where the value of n is incremented for each successive equation. Thus

$$n = 2: \quad \frac{J}{k_2 N_1} = N_1 - \frac{N_2}{(N_1/N^0) K_{1,2}} \tag{2.44}$$

$$n = 3: \quad \frac{J}{k_3 N_1} = N_2 - \frac{N_3}{(N_1/N^0) K_{2,3}} \tag{2.45}$$

$$n = 4: \quad \frac{J}{k_4 N_1} = N_3 - \frac{N_4}{(N_1/N^0) K_{3,4}} \tag{2.46}$$

$$\vdots$$

$$\text{general form}: \frac{J}{k_n N_1} = N_{n-1} - \frac{N_n}{(N_1/N^0) K_{n-1,n}} \tag{2.47}$$

$$\vdots$$

$$n = M: \quad \frac{J}{k_M N_1} = N_{M-1} - \frac{N_M}{(N_1/N^0) K_{M-1,M}} \tag{2.48}$$

We now ask, by what factor must the second equation in the sequence ($n = 3$) be multiplied (or divided) such that, when added to the first equation ($n = 2$), the second term on the right-hand side of the first equation will be canceled by the first term on the right-hand side of the second equation? The answer, by inspection, is that the second equation should be divided by $(N_1/N^0) K_{1,2}$. Performing that division, and adding the first two equations together, yields an equation whose right-hand side is

$$N_1 - \frac{N_2}{(N_1/N^0) K_{1,2}} + \frac{N_2}{(N_1/N^0) K_{1,2}} - \frac{N_3}{(N_1/N^0)^2 K_{1,2} K_{2,3}} = N_1 - \frac{N_3}{(N_1/N^0)^2 K_{1,2} K_{2,3}}. \tag{2.49}$$

We then proceed to the next equation in the sequence ($n = 4$), equation (2.46): By what factor must that equation be divided such that, when added to the equation whose right-hand side is given by equation (2.49), its first right-hand-side term will cancel the last term in the above equation? The answer, again by inspection, is the factor $\left(N_1/N^0\right)^2 K_{1,2} K_{2,3}$.

Comparing this factor to the factor $\left(N_1/N^0\right) K_{1,2}$ used to divide the second equation, it is evident that *all* of the N_n's that appear as the first terms on the right-hand sides of equations (2.44)–(2.48) will be canceled, when the equations are summed together, if all equations in the sequence, beginning with the second (equation 2.45), are divided by a factor $f(n)$ whose general form is given by the following recursion relation:

$$f(2) = 1, \tag{2.50}$$

$$f(n) = f(n-1)\frac{N_1}{N^0}K_{n-2,n-1}, \; n \geq 3. \tag{2.51}$$

For subsequent conciseness, let us define $K_{0,1}$ to be the equilibrium constant for the trivial reaction

$$(0) + A_1 \rightleftarrows A_1. \tag{2.52}$$

As this reaction is the same as reaction (2.29), its free energy change equals zero, and its equilibrium constant equals unity.

Using this definition together with the recursion relation, we conclude that we wish to divide all equations in the sequence (2.44)–(2.48) by the factor

$$\left(\frac{N_1}{N^0}\right)^{n-2} \prod_{i=1}^{n-1} K_{i-1,i}, \; n \geq 2, \tag{2.53}$$

and then sum all of the resulting equations together to produce a single equation. All inner terms on the right-hand side cancel, leaving only the first term of the first equation ($n = 2$) and the second term of the final equation ($n = M$). The result is

$$\sum_{n=2}^{M} \frac{J}{k_n N_1 \left(\frac{N_1}{N^0}\right)^{n-2} \prod_{i=1}^{n-1} K_{i-1,i}} = N_1 - \frac{N_M}{\left(\frac{N_1}{N^0}\right)^{M-1} \prod_{i=1}^{M} K_{i-1,i}}. \tag{2.54}$$

Consider now the term, on each side of this equation, that represents the product of the equilibrium constants for each reaction in the clustering sequence. From equation (2.22),

$$\prod_{i=1}^{n} K_{i-1,i} = \exp\left(-\frac{\Delta G_{0,1}^0 + \Delta G_{1,2}^0 + \Delta G_{2,3}^0 + \cdots + \Delta G_{n-1,n}^0}{k_B T}\right). \tag{2.55}$$

The numerator of the argument of the exponential represents the sum of the stepwise changes in standard Gibbs free energies for the sequence of reactions:

$$(0) + A_1 \rightleftarrows A_1, \; \Delta G_{0,1}^0 = 0 \tag{2.56}$$

$$A_1 + A_1 \rightleftarrows A_2, \; \Delta G_{1,2}^0 \tag{2.57}$$

$$A_2 + A_1 \rightleftarrows A_3, \ \Delta G_{2,3}^0 \qquad (2.58)$$

$$\vdots$$

$$A_{n-1} + A_1 \rightleftarrows A_n, \ \Delta G_{n-1,n}^0 \qquad (2.59)$$

$$\overline{\overline{}}$$

$$n A_1 \rightleftarrows A_n, \Delta G_n^0 \qquad (2.60)$$

As in equation (2.26), the sum of the stepwise changes in standard Gibbs free energy for this sequence of reactions is equal to the standard Gibbs free energy of formation from the monomer vapor of an n-mer. Thus we can write

$$\prod_{i=1}^{n} K_{i-1,i} = \exp\left(-\frac{\Delta G_n^0}{k_B T}\right) = K_n, \qquad (2.61)$$

where K_n denotes the equilibrium constant for the formation of an n-mer from n monomers in the vapor phase. Equation (2.54) can thus be rewritten as

$$\frac{J}{N_1} \sum_{n=2}^{M} \left[k_n \left(\frac{N_1}{N^0}\right)^{n-2} \exp\left(-\frac{\Delta G_n^0}{k_B T}\right) \right]^{-1} = N_1 - \frac{N_M}{\left(N_1/N^0\right)^{M-1} K_M}, \qquad (2.62)$$

where the factor J/N_1 has been removed from the summand as it is independent of the index n.

The term K_M represents the equilibrium constant for the reaction

$$M A_1 \rightleftarrows A_M. \qquad (2.63)$$

Thus

$$K_M = \left[\frac{N_M/N^0}{\left(N_1/N^0\right)^M} \right]_{eq} = \left[\frac{N_M/N_1}{\left(N_1/N^0\right)^{M-1}} \right]_{eq}. \qquad (2.64)$$

Now, M represents some arbitrarily large integer. As a cluster becomes very large, the number density of monomers in equilibrium with it approaches N_s, the vapor number density in equilibrium with the bulk condensed phase at a given temperature. Let us assume that M is large enough so that this approximation is reasonable. We can then replace N_1 in equation (2.64) with N_s and drop the subscript "eq." Substituting this result into equation (2.62) and using $N_1/N_s = S$, the right-hand-side can be written as

$$N_1 - \frac{N_M}{\left(\frac{N_1}{N^0}\right)^{M-1} \left[\frac{N_M/N_s}{\left(N_s/N^0\right)^{M-1}} \right]} = N_1 \left(1 - S^{-M}\right). \qquad (2.65)$$

Then, solving equation (2.62) for the steady-state nucleation rate J, one obtains

$$J = N_1^2 \left(1 - S^{-M}\right) \left\{ \sum_{n=2}^{M} \left[k_n \left(\frac{N_1}{N^0}\right)^{n-2} \exp\left(-\frac{\Delta G_{n-1}^0}{k_B^T}\right) \right]^{-1} \right\}^{-1}. \qquad (2.66)$$

Consider the term $\left(1 - S^{-M}\right)$, where M is arbitrarily large. This term is negative if $S < 1$; zero if $S = 1$; and approaches unity for large values of M for the case $S > 1$.

Thus, for the case $S < 1$, equation (2.66) yields a negative nucleation rate, which makes no physical sense other than as an indication that the system is driven backward toward the state of the monomer vapor. If $S < 1$, there is no tendency for nucleation to occur, so the assumption of a steady-state nucleation current obviously cannot be valid in that case.

For the case $S = 1$, equation (2.66) yields $J = 0$. This does make sense: A perfectly saturated vapor has zero net tendency to nucleate. It can thus be said to experience a steady-state nucleation rate of zero.

Finally, for the case $S > 1$ the term S^{-M} can be neglected in comparison with unity, as M is arbitrarily large, with the result that a positive value is predicted for the steady-state nucleation rate for a supersaturated vapor, given by

$$J = N_1{}^2 \left\{ \sum_{n=2}^{M} \left[k_n \left(\frac{N_1}{N^0}\right)^{n-2} \exp\left(-\frac{\Delta G_{n-1}^0}{k_B T}\right) \right]^{-1} \right\}^{-1}, \quad S > 1. \tag{2.67}$$

The fact that this analysis indicates that a positive nucleation rate requires the saturation ratio to exceed unity is a satisfying result, as it is consistent with what is meant by equilibrium vapor pressure. Subsequently we will omit writing "$S > 1$," as it is understood that single-component steady-state homogeneous nucleation can only occur in this case.

Finally, shifting indices in the summation,[3] and noting that $N_1/N^0 = p_1/p^0$ at a given temperature, one can rewrite this result as

$$J = N_1{}^2 \left\{ \sum_{n=1}^{M} \left[k_{n+1} \left(\frac{p_1}{p^0}\right)^{n-1} \exp\left(-\frac{\Delta G_n^0}{k_B T}\right) \right]^{-1} \right\}^{-1}. \tag{2.68}$$

We will refer to this result as the "summation expression" for the steady-state nucleation rate. To proceed further, one requires size-dependent values for the forward rate constants and Gibbs free energies of cluster formation, if not up to the "arbitrarily large" size M, then at least up to a large enough value of n for the summation to have converged. In the next chapter, we apply classical theory to evaluate these terms, which allows the above summation expression to be converted to an algebraic expression for the nucleation rate. In Chapter 6, we return to the summation expression and discuss the direct use of atomistic values for k_n and ΔG_n^0. Convergence of the summation and the relation between convergence and the critical cluster size are discussed in Section 6.3.

2.7   Effect of Pressure on Gibbs Free Energy

The superscript "0" in ΔG_n^0 denotes the free energy of cluster formation at standard pressure, conventionally 1 atm or 1 bar. Certainly that is the most useful form if tables

---

[3] As M is arbitrarily large, we retain it as the upper limit of n rather than replacing it by $M - 1$.

are available of thermodynamic properties for clusters of interest. However, presenting the nucleation rate in the form of equation (2.68) obscures the effect of saturation ratio, other than that it implicitly assumes $S > 1$ based on equation (2.66).

The N_1^2 term in front of the summation term is equivalent to $N_s^2 S^2$. Indeed, as discussed in Section 3.11, experimental studies of nucleation rate typically present results in terms of $J(S)$ along isotherms, in which case comparisons with theory require this substitution, or its equivalent in terms of partial pressures. However, if one wishes to understand the effect of saturation ratio on nucleation rate, this substitution is misleading. The reason for this is that the prefactor N_1^2 arises from purely kinetic not thermodynamic considerations. Namely, from equation (2.12),

$$R_2 = k_2 N_1^2, \tag{2.69}$$

that is, the forward rate of dimerization scales on the square of the monomer number density, and the nucleation rate scales to some extent on R_2, as it is the first step in the clustering sequence. Saturation ratio is a thermodynamic concept that characterizes a vapor's degree of nonequilibrium with respect to the bulk condensed phase. However, the forward rate of dimerization is insensitive to this degree of nonequilibrium and scales only on N_1^2, regardless of the saturation ratio. Nevertheless, the effect of thermodynamics, as embedded in the saturation ratio, is indeed crucial for the nucleation rate. To see this, we need to consider the effect of pressure on Gibbs free energy.

Consider the definition of Gibbs free energy, equation (1.1), $G \equiv H - TS$, where S here is entropy. Thus, for a system that undergoes an infinitesimal change of state, one has

$$dG = dH - TdS - SdT. \tag{2.70}$$

If temperature is fixed, this reduces to

$$dG = dH - TdS. \tag{2.71}$$

From the Gibbs equation for a simple compressible substance,

$$TdS = dH - Vdp, \tag{2.72}$$

where V is total volume. Combining these last two equations gives

$$dG = Vdp \tag{2.73}$$

for the change in Gibbs free energy in an isothermal process.

2.7.1   Clusters as Ideal Gases

As stated in Section 2.2, we assume here that all n-mers behave as ideal gases and thus obey the ideal gas equation of state, equation (2.15). In that case, one can write

$$V = \frac{N_{tot} k_B T}{p}, \tag{2.74}$$

where N_{tot} is the total number of molecules in the system.[4] Using this expression in equation (2.73), for an ideal gas of fixed chemical composition undergoing an isothermal process, the change in free energy as pressure changes from p_A to p_B is given by[5]

$$[G(p_B) - G(p_A)]_{\text{system}} = N_{tot} k_B T \int_{p_A}^{p_B} \frac{dp}{p}. \qquad (2.75)$$

Carrying out this integration and dividing by N_{tot}, one has the effect of pressure on Gibbs free energy, on a per-molecule basis:

$$G(p_B) - G(p_A) = k_B T \ln \frac{p_B}{p_A}. \qquad (2.76)$$

Thus, for example,

$$G(p_1) = G(p_s) + k_B T \ln S, \qquad (2.77)$$

where we have used equation (1.7) for the definition of saturation ratio.

Therefore, as ΔG_n is the free energy change for reaction (2.25), one finds that

$$\Delta G_n(p_1) = \Delta G_n^0 - (n-1) k_B T \ln \frac{p_1}{p^0}, \qquad (2.78)$$

or, equivalently,

$$\Delta G_n(p_1) = \Delta G_n(p_s) - (n-1) k_B T \ln S. \qquad (2.79)$$

Thus equation (2.68) can be rewritten as either

$$J = N_1^2 \left\{ \sum_{n=1}^{M} \left[k_{n+1} \exp\left(-\frac{\Delta G_n(p_1)}{k_B T} \right) \right]^{-1} \right\}^{-1} \qquad (2.80)$$

or

$$J = N_1^2 \left\{ \sum_{n=1}^{M} \left[k_{n+1} S^{n-1} \exp\left(-\frac{\Delta G_n(p_s)}{k_B T} \right) \right]^{-1} \right\}^{-1}. \qquad (2.81)$$

Alternatively, one could have obtained these two equations simply by setting the arbitrary standard pressure in equation (2.68) to equal either p_1 or p_s, respectively. Each of these expressions proves useful in Chapters 3 and 6. Often in the nucleation literature, one finds the simple notation "ΔG," with it usually being assumed that it is evaluated at the actual partial pressure p_1 of the monomer vapor, while in the thermochemical literature, "ΔG^0," with p^0 taken as 1 bar or 1 atm, is commonly found.

---

[4] Here "molecules" is broadly defined to include all freely translating entities. Thus, in this context, a cluster is a single molecule, regardless of the number of molecules of which it is composed, and $N = N_{tot}/V$ is the total molecular number density.

[5] While G is a function of both temperature and pressure, for economy of notation we shall usually emphasize the pressure dependence, with the fact that G also depends on temperature being understood. Thus $G(p)$ should be understood to stand for $G(T, p)$ unless stated otherwise.

In any case, comparing these three equivalent expressions for the nucleation rate, equations (2.68), (2.80), and (2.81), the most concise expression is equation (2.80), which evaluates ΔG_n at the actual partial pressure p_1 of the monomer vapor. Then equation (2.81), which arises purely from thermodynamics, clearly shows the effect of saturation ratio on $\Delta G_n(p_1)$.

Equation (2.76) can also be used to recast the stepwise free energy change from any pressure p_A to any other pressure p_B. As the stepwise free energy change applies to reaction (2.4), equation (2.76) yields

$$\Delta G_{n-1,n}(p_B) = \Delta G_{n-1,n}(p_A) - k_B T \ln \frac{p_B}{p_A}. \tag{2.82}$$

2.7.2   Clusters as Incompressible Substances

The classical theory of nucleation, discussed in Chapter 3, treats individual clusters as liquid droplets that have the same properties as the bulk liquid. Under this assumption, individual clusters can be modeled as incompressible substances.

An incompressible substance is a limiting case of a simple compressible substance, in which volume change can be neglected. Thus, equation (2.73) still applies,

$$dG = VdP, \tag{2.73}$$

but with volume now treated as constant rather than obeying the ideal gas equation of state. For a finite change of pressure from p_A to p_B, the change in Gibbs free energy is then given by

$$G(p_B) - G(p_A) = \int_{p_A}^{p_B} Vdp = V(p_B - p_A). \tag{2.83}$$

2.8   Nucleation Theorem

As discussed in Chapter 1, if the saturation ratio is greater than unity but not too high, then one is in the condensation/evaporation regime, where a critical size n^* exists such that clusters smaller than n^* tend on average to decay, while larger clusters tend to grow, consistent with a system's tendency to seek the minimum in its total Gibbs free energy. From equation (2.79), it follows that $\Delta G_n(p_s)$ is generally positive and that the $(n-1)k_B T \ln S$ term causes a maximum in $\Delta G_n(p_1)$ to exist at some finite cluster size n^*. But if $\Delta G_n(p_1)$ has a maximum, then the fact that it is exponentiated in equation (2.80) means that the summand for which n equals n^* represents a much stronger maximum. Indeed, it turns out that terms close to the critical size typically dominate the summation, and under some circumstances, the summation may be dominated by the single term associated with this maximum.

For now, let us assume this to be true. With this "one-term approximation," one can write, following equation (2.81),

$$J_{1-\text{term}} \approx CN_1{}^2 k_{n^*+1} S^{n^*-1} \exp\left[-\frac{\Delta G^*(p_s)}{k_B T}\right], \tag{2.84}$$

where ΔG^* denotes ΔG_{n^*} and C is a dimensionless factor, not too much larger than unity, that accounts for the fact that other terms also contribute to the summation.[6] Noting that $N_1 = SN_s$, and taking the logarithm of both sides,

$$\ln J \approx \ln\left\{C k_{n^*+1}[N_s(T)]^2\right\} - \frac{\Delta G^*(p_s)}{k_B T} + (n^* + 1)\ln S. \tag{2.85}$$

The condensation rate constant depends on temperature but not on saturation ratio. Thus, assuming that the factor C is at most a weak function of S, for a given temperature one has the result

$$\left(\frac{\partial \ln J}{\partial \ln S}\right)_T = n^* + 1. \tag{2.86}$$

This relation is known as the nucleation theorem, or sometimes the first nucleation theorem. The nucleation theorem states that the dependence of the steady-state nucleation rate on saturation ratio, at given temperature, has a simple relation to the value of n^*. Conversely, equation (2.86) can be rewritten as

$$n^* = \left(\frac{\partial \ln J}{\partial \ln S}\right)_T - 1. \tag{2.87}$$

Thus, if one has experimental measurements of the nucleation rate for a range of saturation ratios at fixed temperature, the nucleation theorem allows one to infer the critical size.

Note that the derivation of the nucleation theorem is independent of the functional form of $\Delta G(p_s)$ and thus is valid for both classical and atomistic approaches. For more rigorous derivations, see Kashchiev (1982) and Ford (1997).

Homework Problems

2.1 Derive equations (2.79) and (2.82).

2.2 Consider some function $F(n)$ that is given by

$$F(n) = A(n-1) - B(n), \ 2 \le n \le 5,$$

where A and B are arbitrary functions and n is an integer. Suppose that at steady state F becomes independent of n. Following a similar approach to that in Section 2.6 for the derivation of the summation expression for the steady-state nucleation rate, develop an expression for $F(A, B)$ at steady state.

2.3 Show that the steady-state nucleation rate given by equation (2.68) is not affected by the choice of standard pressure.

---

[6] As will be seen in Chapters 3 and 6, C can become significantly larger than unity only for saturation ratios close to unity, for which nucleation rates are generally quite small.

3 Classical Nucleation Theory
The Liquid Droplet Model

3.1   Classical versus Atomistic Theories

In the previous chapter, we derived an expression for the steady-state rate of single-component nucleation from a supersaturated vapor. This expression requires knowledge of the values of the rate constants k_n for monomer condensation and especially, because of the exponential term in equation (2.68), for the Gibbs free energies of cluster formation, ΔG_n. These quantities are required as a function of cluster size, up to a size at least large enough for the summation in equation (2.68) to have converged. As we shall see, this size is usually associated with an entity known as the "critical cluster."

Theories of particle nucleation from the gas phase can be divided into "classical" and "atomistic" (or "nonclassical") theories. In general, classical theories model clusters as being very small versions of the bulk condensed phase of the same substance. More specifically, classical nucleation theory models clusters as structureless, spherical liquid droplets. In other words, while from the historical viewpoint the term "classical" refers to the original development of the theory, in the late nineteenth century and first half of the twentieth century, it is appropriately used to mean any theory that models microscopic clusters in macroscopic terms. Atomistic theories, in contrast, account for the fact that the critical-size nuclei and smaller clusters are usually composed of too few atoms or molecules to be well described in terms of macroscopic properties. These theories attempt to utilize the structures, properties, and reactivities of chemically specific clusters, at an atomic/molecular level. These structures and properties may differ drastically for different-sized clusters of the same substance, or even for the same size and stoichiometry, for different isomers. Moreover, even though we have referred to monomer addition and elimination reactions as "condensation" and "evaporation" reactions, phase change *per se* does not explicitly enter into an atomistic theory, in contrast to the approach of classical theory.

The origins of classical nucleation theory (CNT) can be traced to Gibbs, who in 1876–1878 developed a theory "On the Possibility of Formation of a Fluid of Different Phase within any Homogeneous Fluid" (Bumstead and Van Name 1906). Subsequently, from the 1920s to the 1940s, a detailed theory was developed that yielded an analytical expression for the steady-state rate of single-component nucleation from a supersaturated vapor, with key contributions by Volmer and Weber (1926), Farkas (1927),

Becker and Döring (1935), and Zeldovich (1943). The theory is commonly referred to as the "Becker–Döring theory," or the "Becker–Döring–Zeldovich theory."

Given the approximate, macroscopic approach of CNT, it may seem surprising that it should now be of any more than historical interest, but in fact it is still the basis of most practical homogeneous nucleation calculations in the literature. This situation will presumably change as more atomistic data on clusters become available. In any case, consideration of the classical model provides considerable insight into the problem. Classical nucleation theory does not, in general, do a good job of predicting the quantitative value of the nucleation rate, but it is still often used to provide estimates of nucleation rate, lacking better information.

3.2 Clusters as Liquid Droplets

Referring to the forward and backward directions of reaction (2.9) as "condensation" and "evaporation" events implies that the cluster represents a condensed phase. While this makes little sense for very small clusters, it does suggest a simple physical model. Gibbs, in 1876–1878, had the idea that a reasonable model for a small cluster in a system undergoing gas-to-liquid phase change, lacking better information, was to approximate it as a spherical liquid droplet (Bumstead and Van Name 1906). This droplet is assumed to have the following properties:

(1)   It has the same mass density as the bulk liquid of the same substance at the same temperature T.
(2)   It has the same surface tension as a flat liquid surface of the same substance in equilibrium with its vapor at temperature T. This assumption is known as the "capillarity" approximation.

A key feature of this model is that the cluster is considered to be of a condensed phase – specifically, liquid – regardless of how small it is. Even though this model may deviate far from reality, it has proven to be exceptionally robust.

As an example, consider a 10-molecule water cluster, $(H_2O)_{10}$. Figure 3.1 shows the prediction of ab initio quantum chemical calculations for the most stable $(H_2O)_{10}$ isomers (Maheshwary et al. 2001). Obviously, the clusters shown have structure and are far from being spherical. However, CNT models $(H_2O)_{10}$ as being a structureless sphere with a diameter of 0.829 nm. This calculation follows from Section 1.2.2. The effective diameter of a water molecule, based on the density of bulk liquid water at 25°C and 1 atm, can be written as $d_1 = 0.385$ nm. Therefore, assuming that the droplet volume scales directly on the number of molecules, the effective diameter of $(H_2O)_{10}$ in the classical model is $d_{10} = (0.385 \text{ nm}) \times 10^{1/3} = 0.829$ nm.[1]

---

[1] This result is affected by temperature and pressure, although only slightly, as liquid water is close to incompressible. In any case, it is obviously an approximation.

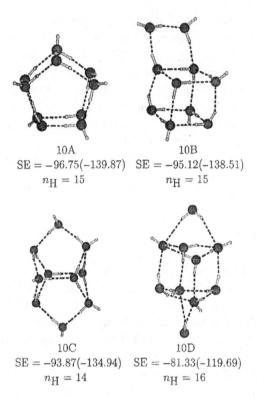

10A

$\text{SE} = -96.75(-139.87)$

$n_H = 15$

10B

$\text{SE} = -95.12(-138.51)$

$n_H = 15$

10C

$\text{SE} = -93.87(-134.94)$

$n_H = 14$

10D

$\text{SE} = -81.33(-119.69)$

$n_H = 16$

Figure 3.1 Structure of most stable $(H_2O)_{10}$ clusters, predicted by ab initio calculations. Reprinted with permission from Maheshwary et al. (2001). Copyright 2001 American Chemical Society.

3.3   The Classical Model for the Thermodynamics of Cluster Formation

The classical model provides a means for estimating the Gibbs free energy of formation of an n-mer. In addition to facilitating evaluation of equation (2.80), this estimate affords useful qualitative insight into the thermodynamics of clustering.

3.3.1   Division of Droplet Free Energy into Volume and Surface Contributions

To evaluate the Gibbs free energy of cluster formation in the classical model, we first develop an expression for the stepwise free energy change associated with the growth of a cluster from size $(n-1)$ to size n via monomer condensation. As given by equation (2.26), the Gibbs free energy of n-mer formation is simply equal to the sum of these $(n-1)$ stepwise changes.

For the condensation reaction,

$$A_{n-1} + A_1 \rightleftarrows A_n, \ n \geq 2, \tag{2.9}$$

evaluated at pressure p_1, the change in free energy can be written as

$$\Delta G_{n-1,n}(p_1) = G_n(p_1) - G_{n-1}(p_1) - \tilde{G}_v(p_1), \; n \geq 2. \tag{3.1}$$

Now, from the classical viewpoint, the free energy of an n-mer droplet can be expressed as the sum of two contributions, a volume contribution and a surface (or interface) contribution:

$$G_n = G_{n,\text{volume}} + G_{n,\text{surface}}. \tag{3.2}$$

Gibbs free energy is an extensive thermodynamic property of a system, so a system consisting of n molecules of liquid, as the liquid droplet model assumes for an n-mer, has an associated volume free energy that can be written as

$$G_{n,\text{volume}} = n\tilde{G}_l, \tag{3.3}$$

where $\tilde{G}_l$ is the free energy per molecule of the bulk liquid. Thus the change in a cluster's volume free energy associated with condensation of a single monomer is simply $\tilde{G}_l$.

3.3.2 The Free Energy Change Associated with Stretching a Surface

For a liquid droplet, one must in addition account for the free energy associated with its surface. That is because work is required to stretch a surface, and as a droplet grows, its surface expands. Surface tension σ is defined such that stretching a surface so as to increase its area s by an infinitesimal amount ds requires reversible work in the amount.

$$\delta W = \sigma ds. \tag{3.4}$$

Thus σ has units of $J \cdot m^{-2}$ or, equivalently, $N \cdot m^{-1}$. When a droplet's surface expands, its volume increases, and one must also account for the resulting work done by the droplet against its surroundings,

$$\delta W = pdV, \tag{3.5}$$

where V is the droplet volume. Combining these two expressions, the work done *by* a droplet *against* its surroundings when its surface stretches is given by

$$\delta W = pdV - \sigma ds, \tag{3.6}$$

where we have accounted for the fact that the surface tension term represents work done on the droplet, and thus is negative.

Consider a closed system undergoing an infinitesimal change of state during a reversible process at fixed temperature and pressure. By the first law of thermodynamics, the change in the system's internal energy, assuming that kinetic and potential energy effects are negligible, can be written as

$$dE = \delta Q - \delta W, \tag{3.7}$$

where Q represents heat transfer to the system and W work done by the system. The system's change in entropy for the reversible process can be written as

$$dS = \frac{\delta Q}{T}, \tag{3.8}$$

where S in this context is entropy.

Combining these expressions gives

$$TdS = dE + \delta W. \tag{3.9}$$

Substituting equation (3.6) for the work, we have

$$TdS = dE + pdV - \sigma ds. \tag{3.10}$$

This is the Gibbs equation for a substance whose surface can be reversibly stretched. In the absence of surface stretching, it reduces to the familiar Gibbs equation for a simple compressible substance.

Using the definition of enthalpy, equation (1.2), equation (3.10) can be rewritten as

$$dH - TdS = Vdp + \sigma ds. \tag{3.11}$$

Then using the definition of Gibbs free energy, equation (1.1), equation (3.11) can be converted to

$$dG = Vdp - SdT + \sigma ds. \tag{3.12}$$

At given temperature and pressure, one thus has the result

$$dG = \sigma ds \tag{3.13}$$

for the increase in free energy associated with stretching a droplet surface by an amount ds.

In the liquid droplet model, an increase in a cluster's surface area by a finite amount Δs requires reversible work in the amount $\int_{\Delta s} \sigma ds$ and hence contributes that amount to the change in the cluster's Gibbs free energy. For very small liquid droplets, surface tension can vary with droplet size. However, most versions of CNT make the capillarity approximation, meaning that they assume that σ is independent of cluster size, in which case the contribution is simply $\sigma \Delta s$. In that case, condensation of a monomer on an $(n-1)$-mer results in an increase in surface free energy in the amount $\sigma(s_n - s_{n-1})$.

3.3.3   The Stepwise Free Energy Change in the Classical Model

Considering, then, both volume and surface contributions to G_n, equation (3.1) can be rewritten as

$$\Delta G_{n-1,n}(p_1) = \tilde{G}_l(p_1) - \tilde{G}_v(p_1) + \sigma(s_n - s_{n-1}), n \geq 2. \tag{3.14}$$

Now, assuming that the vapor is an ideal gas and that the droplet is a pure liquid, the free energies per molecule are identical to the chemical potentials $\tilde{\mu}$ per molecule in the respective phases, both evaluated in this case at pressure p_1. Thus equation (3.14) can be rewritten as

$$\Delta G_{n-1,n}(p_1) = \tilde{\mu}_l(p_1) - \tilde{\mu}_v(p_1) + \sigma(s_n - s_{n-1}), n \geq 2. \tag{3.15}$$

The advantage of switching from Gibbs free energy to chemical potential is that it allows us to exploit the fact that a fundamental condition of equilibrium between two phases of the same substance is that their chemical potentials must be equal. As $p_s(T)$ is the equilibrium vapor pressure above a flat surface of a pure liquid, it follows that

$$\tilde{\mu}_l(p_s, T) = \tilde{\mu}_v(p_s, T). \tag{3.16}$$

We thus wish to relate the chemical potentials of the liquid and the vapor evaluated at p_1 to their values at p_s.

For the vapor, which we assume behaves as an ideal gas, by equation (2.77) we have

$$\tilde{\mu}_v(p_1, T) = \tilde{\mu}_v(p_s, T) + k_B T \ln \frac{p_1}{p_s}. \tag{3.17}$$

For the liquid, which is modeled as an incompressible substance, we instead have equation (2.83), which here can be written as

$$\tilde{\mu}_l(p_1, T) = \tilde{\mu}_l(p_s, T) + \tilde{v}_l(p_1 - p_s), \tag{3.18}$$

where $\tilde{v}_l$ is the molecular volume of the liquid.

Using these relations, and utilizing equation (3.16), equation (3.15) for the stepwise free energy change becomes

$$\Delta G_{n-1,n}(p_1) = \tilde{v}_l(p_1 - p_s) - k_B T \ln \frac{p_1}{p_s} + \sigma(s_n - s_{n-1}), \, n \geq 2, \tag{3.19}$$

or, using the definition of saturation ratio,

$$\Delta G_{n-1,n}(p_1) = p_s \tilde{v}_l(S - 1) - k_B T \ln S + \sigma(s_n - s_{n-1}), \, n \geq 2. \tag{3.20}$$

In most situations of interest, the liquid-phase pressure correction term on the right hand of this equation is negligible compared to the gas-phase pressure correction term.[2] That is,

$$p_s \tilde{v}_l(S - 1) \ll k_B T \ln S. \tag{3.21}$$

Therefore, assuming that the first term on the right-hand side of equation (3.20) can be neglected in comparison with the second term, and rearranging, one finally has the result that the stepwise change in Gibbs free energy in CNT is given by

$$\Delta G_{n-1,n}(p_1) = \sigma(s_n - s_{n-1}) - k_B T \ln S, \, n \geq 2. \tag{3.22}$$

It is interesting to note that, for the special case $p_1 = p_s$, the stepwise free energy change reduces to

$$\Delta G_{n-1,n}(p_s) = \sigma(s_n - s_{n-1}), \, n \geq 2. \tag{3.23}$$

This relation can be said to convey the essence of the classical model, as the $k_B T \ln S$ term in equation (3.22) can be regarded as simply a correction associated with the effect of pressure on the entropy of an ideal gas. Strictly speaking, the pressure-volume product, that is, the first term on the right-hand side of equation (3.20), is also an

---

[2] See homework problem 3.1.

intrinsic aspect of the classical model, but because it is very small compared to other terms, it is usually neglected in CNT.

3.3.4 The Gibbs Free Energy of Cluster Formation

The Gibbs free energy of cluster formation in CNT can now be evaluated as in equation (2.26) by summing equation (3.22) over all the monomer condensation events that form an n-mer from n vapor-phase monomers:

$$\Delta G_n(p_1) = \sum_{i=2}^{n} \Delta G_{i-1,i}(p_1), \; n \geq 2. \tag{3.24}$$

Substituting the right-hand side of equation (3.22) for the summand yields

$$\Delta G_n(p_1) = \sigma(s_n - s_1) - (n-1)k_B T \ln S, \; n \geq 2, \tag{3.25}$$

for the free energy of formation of an n-mer in the classical model, evaluated at the actual monomer partial pressure p_1.

As our derivation of this equation for ΔG_n was based on summing over all the $(n-1)$ stepwise free energy changes in a cluster's growth from a monomer to an n-mer, it follows that it applies only for $n \geq 2$. Indeed, there is no need to model ΔG_1, as, by equation (2.30), it identically equals zero. Nevertheless, it can be noticed that if one applies equation (3.25) to the monomer, one does in fact obtain $\Delta G_1 = 0$. Thus the qualifier "$n \geq 2$" can be removed from equation (3.25).

In contrast, the standard development of CNT, rather than proceeding through the stepwise reactions involved in growing an n-mer, goes directly to a model for an n-mer droplet and thereby neglects to subtract s_1 from s_n in the first term on the right-hand side of equation (3.25), and also neglects to subtract 1 from n in the second term. These omissions result in various inconsistencies in the theory and finally make a rather large difference in the result for the steady-state nucleation rate. The approach followed here is known as the "self-consistent classical theory" or the "internally consistent classical theory" (Blander and Katz 1972; Girshick and Chiu 1990).[3] A further discussion of the differences between the "standard" and "self-consistent" theories is given in Section 3.10.

Under the assumption that clusters are spheres of uniform density, one can write

$$s_n = n^{2/3} s_1. \tag{3.26}$$

Dividing both sides of equation (3.25) by $k_B T$, a dimensionless free energy of formation can then be written as

$$\frac{\Delta G_n(S,T)}{k_B T} = \left(n^{2/3} - 1 \right) \Theta - (n-1) \ln S, \tag{3.27}$$

where Θ is a dimensionless surface tension, defined by

---

[3] We use the acronym "CNT" to refer to classical nucleation theory in general, including both the self-consistent version and the "standard" version.

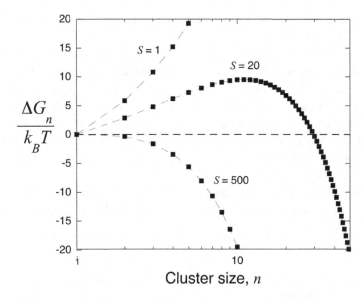

Figure 3.2 Gibbs free energy of cluster formation versus cluster size, in self-consistent classical model, for various saturation ratios and for a value of the dimensionless surface tension $\Theta = 10$.

$$\Theta = \frac{\sigma S_1}{k_B T}. \tag{3.28}$$

Values of Θ for a wide variety of liquid substances and realistic temperatures range from about 3 to 50. For example, for water at 25°C, $\Theta = 8.16$.

Equation (3.27) indicates that the free energy of cluster formation equals the difference between two terms, one of which scales on cluster surface area ($\propto n^{2/3}$) and the other of which scales on cluster volume ($\propto n$). Equation (3.27) is plotted in Figure 3.2 for various values of S and for $\Theta = 10$.

3.4 Limiting Cases

Figure 3.2 can be interpreted as follows. Consider a closed system at fixed temperature and pressure that consists solely of a monomer vapor and of a single cluster of the same substance. Then ΔG_n represents the total free energy of this system, for various sizes of the cluster, relative to a system that consists of the same number of molecules,[4] all in the vapor phase. When the cluster adsorbs or desorbs a monomer, the total free energy of the system either increases or decreases, as indicated in Figure 3.2.

Now consider a closed system that consists of a monomer vapor and a large population of clusters of the same substance. The clusters can grow or decay by condensation/evaporation events, but are assumed not to interact with each other. Then the total free energy of this system, relative to a system that consists of the same number of

---

[4] Here we include the number of molecules of which the cluster is comprised.

molecules in the vapor phase, is given simply by the sum of all the individual cluster ΔG_n's.

Regardless of the state of the system, some clusters can grow while others shrink, through discrete condensation/evaporation events. However, thermodynamics says that a closed system at fixed temperature and total pressure will adjust its overall configuration until it eventually reaches the minimum possible value in its total Gibbs free energy, consistent with the total mass of the system, the specified elemental composition, and the temperature and pressure. Therefore, the net result of all the individual cluster growth/decay events must be in the direction that reduces the system's Gibbs free energy.

From Figure 3.2, one can note three distinct regimes: (1) a perfectly saturated vapor, $S = 1$, for which ΔG_n is a monotonically increasing function of n given by

$$\frac{\Delta G_n(p_s, T)}{k_B T} = \left(n^{2/3} - 1\right)\Theta; \tag{3.29}$$

(2) very large saturation ratios, for which ΔG_n is a monotonically decreasing function of n; and (3) intermediate saturation ratios, for which ΔG_n has a maximum at some cluster size $n > 1$.

3.4.1   Perfectly Saturated Vapor

For the case of a perfectly saturated vapor, $S = 1$, the $\ln S$ term in equation (3.27) vanishes, and the free energy of formation of each cluster is a monotonically increasing function of cluster size that scales on the cluster surface area. Thus all increases in cluster size are accompanied by an increase in the system's Gibbs free energy. Therefore, clusters of any size above the monomer tend, on average, to shrink. As cluster size increases, so does the slope of the free energy curve, posing a strong thermodynamic barrier to the growth of clusters to large sizes.

3.4.2   Very Large Saturation Ratios

For saturation ratios higher than some value S^*, the value of ΔG_n for all clusters larger than the monomer is negative and monotonically decreases, because the volume term in equation (3.27) is larger than the surface area term, even for the dimer. From equation (3.27), one finds that

$$S^* = \exp\left[\left(2^{2/3} - 1\right)\Theta\right] \approx e^{0.587\Theta}. \tag{3.30}$$

For example, for $\Theta = 10$, one finds $S^* = 356$. Thus in Figure 3.2 the case of $S = 500$ is seen to lie in this regime.

In this regime, it is thermodynamically favorable for monomers and clusters of all sizes to grow. This condition corresponds to the "collision-controlled regime" discussed in Section 1.2.3. The stable equilibrium is found at $n \rightarrow \infty$, that is, the bulk condensed phase.

3.5   Intermediate Saturation Ratios: The Condensation/Evaporation Regime

The intermediate values of the saturation ratio, $1 < S < S^*$, correspond to the "condensation/evaporation regime" discussed in Chapter 1. Here the free energy of cluster formation has a positive maximum at some cluster size $n^* > 1$, as seen in Figure 3.2 for the case of $S = 20$. Consistent with the thermodynamic imperative to reduce the total free energy of the system, clusters that are smaller than n^* tend to shrink, while clusters larger than n^* tend to grow. Again, these tendencies are only averages over a large population of clusters. Individual clusters experience discrete condensation/evaporation events that occur in random order, and so an individual cluster may, temporarily at least, climb uphill in ΔG_n-space, against the thermodynamic driving force. Therefore it is possible that some clusters, initially smaller than size n^*, may succeed in growing to n^*, despite the fact that most of the other clusters in their initial size cohort have shrunk (assuming that their initial size was larger than the monomer).

3.5.1   Unstable Equilibrium

If we temporarily ignore the fact that n, which represents the number of molecules in a cluster, is inherently an integer, and thus convert the discrete function ΔG_n to a continuous function $\Delta G(x)$ of a continuous size variable x, then the location of the maximum in $\Delta G(x)$ occurs at a value $x = x^*$ for which

$$\left. \frac{\partial(\Delta G_x)}{\partial x} \right|_{T,S;\, x=x^*} = 0. \tag{3.31}$$

Indeed that is the condition for equilibrium of the system, as it represents a state in which there is no tendency for the total Gibbs free energy of the system either to increase or decrease as the cluster size changes. However, the equilibrium here is not stable, because if a cluster of size x^* should grow or shrink by an infinitesimal amount, then returning to x^* would require an increase in the Gibbs free energy of the system, whereas the laws of thermodynamics require that the system will continually adjust its configuration in the direction of decreasing Gibbs free energy, until finally a minimum value is reached. Precisely at $x = x^*$ there is no tendency for the cluster to shrink or to grow. However, if $x < x^*$ the cluster is likely to shrink, while if $x > x^*$, the cluster is likely to grow.

3.5.2   The Critical Cluster Size

Evidently, for a cluster to grow from the size of a monomer to the size x^*, it must climb uphill in free energy space, against the direction preferred by thermodynamics. However, once the size x^* is reached and then only infinitesimally exceeded, thermodynamics indicates that it is likely to grow. Therefore, $\Delta G(x^*)$ is referred to as the "free energy barrier for nucleation," and x^* is referred to as the "critical cluster size," or the size of the "critical nucleus." It is thus evident, from thermodynamics alone, that the problem of single-component homogeneous nucleation is equivalent to the problem of finding the rate of formation of these nuclei, which requires overcoming the free energy barrier $\Delta G(x^*)$. It is further evident, from the viewpoint of kinetics, that the existence of a critical size represents a rate-limiting bottleneck to nucleation.

To find the value of x^* in CNT, one needs only to evaluate equation (3.31), substituting the continuous variable x for the integer number of molecules n in equation (3.27). One obtains

$$x^* = \left(\frac{2}{3}\frac{\Theta}{\ln S}\right)^3,$$
(3.32)

which is known as the Gibbs–Thomson equation. For example, for $\Theta = 10$ and $S = 10$, one finds $x^* = 24.3$.

Note that the actual critical size n^* is an integer representing the value of n at which ΔG_n has its maximum value. In most cases in CNT, this corresponds to the nearest integer obtained by rounding the value of x^*. However, it can be noted that $\Delta G_x(p_1)$ is slightly asymmetric about its maximum at $x = x^*$, rising more steeply for $x < x^*$ than it falls for $x > x^*$. This implies that in cases where x^* lies precisely midway between two integers, n^* will equal the lower integer value, and this relation persists even for values of x^* that are slightly larger than the midpoint between two integers.

In Chapter 1, the critical cluster is referred to as an "embryo," consistent with the fact that it is the size beyond which the cluster is likely to grow to become a condensed-phase "particle." It is worth remarking that CNT has engendered some confusion around this point, as CNT treats all clusters, including subcritical ones, as liquid droplets. But in any atomistic model, critical clusters may have structure, and thus may be termed "solid-like," or may be structureless, and thus may be termed "liquid-like," but, following the discussion in Section 1.2.2, that does not necessarily qualify them to be considered as condensed-phase particles.

3.5.3   The Kelvin Equation

Figure 3.2 lends itself to another interpretation. For a cluster of given size x, there exists a unique value of the saturation ratio such that x exactly equals the size of the critical nucleus, which has no tendency either to grow or to decay. This value of S, shown in Figure 3.3 by the locus of maxima in the free energies of cluster formation, can thus be interpreted as the saturation ratio that exists in equilibrium with a cluster of size x. Solving equation (3.32) for S gives

$$S = \exp\left(\frac{2\Theta}{3n^{2/3}}\right),$$
(3.33)

for the saturation ratio of vapor existing in equilibrium with a cluster containing n monomers.

Converting equation (3.32) from the continuous-number-of-monomers function x to the continuous cluster diameter function d_x, one can rewrite equation (3.32) to give the diameter of the critical nucleus:

$$d_x^* = \frac{4\sigma v_1}{k_B T \ln S}.$$
(3.34)

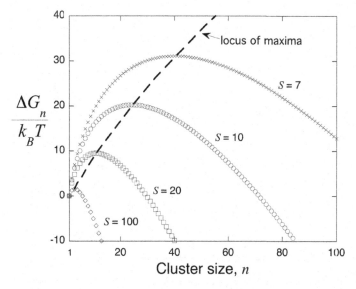

Figure 3.3 Gibbs free energy of cluster formation versus cluster size for various saturation ratios and for a value of the dimensionless surface tension $\Theta = 10$. The locus of maxima gives the value of the free energy of cluster formation at the critical size for each saturation ratio, as well as the value of the saturation ratio for vapor coexisting in equilibrium with clusters of each size.

Solving this for $S = p_1/p_s$, one obtains

$$p_1 = p_s \exp\left(\frac{4\sigma v_1}{k_B T d}\right), \tag{3.35}$$

where p_1 can now be interpreted as the equilibrium vapor pressure over a droplet of diameter d. This result is known as the Kelvin equation.

A physical explanation for the qualitative trend given by equation (3.35) can be given as follows. Consider a molecule on the surface of a droplet. As the droplet shrinks, its radius of curvature decreases, and the number of molecules immediately adjacent to the molecule of interest decreases. The adjacent molecules exert attractive forces that restrain the molecule from evaporating. Thus, as the droplet shrinks, the molecule of interest becomes less strongly bound to the droplet, that is, more likely to evaporate, with the result that the equilibrium vapor pressure is higher.

Figure 3.4 graphs equation (3.35), assuming bulk properties of liquid water at 25°C. The deviation of the equilibrium vapor pressure from the value p_s for equilibrium above an infinite flat surface is seen to be small for a 100-nm-diameter water droplet, but becomes increasingly significant as droplet size decreases. For a 5-nm-diameter water droplet, the Kelvin equation predicts a 52% increase in the equilibrium vapor pressure compared to p_s. A 5-nm-diameter water droplet contains approximately 2,500 molecules, suggesting that the spherical droplet approximation, on which the Kelvin equation is based, may be reasonable. For a 1-nm droplet, the predicted equilibrium vapor pressure is more than 8 times higher than p_s. However, a 1-nm-diameter water droplet contains only about 20 molecules, which makes the validity of the classical droplet model questionable.

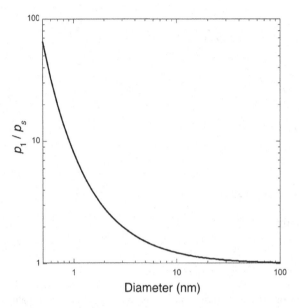

Figure 3.4 Ratio of equilibrium vapor pressure over droplet of finite diameter to equilibrium vapor pressure over infinite flat surface, as given by equation (3.35) for water at 25°C.

3.6   Rate Constants in Classical Theory

3.6.1   Condensation Rate

The forward rate for monomer addition depends on the rate of collisions between monomers and n-mers, and on the fraction of those collisions in which the monomer sticks to the n-mer. Thus one can write

$$R_n = k_n N_{n-1} N_1 = \alpha_{n-1} Z_{1,n-1}, \; n \geq 2, \tag{3.36}$$

where α_{n-1} is the "sticking coefficient" (or "mass accommodation coefficient") of a monomer onto an $(n-1)$-mer, and $Z_{1,n-1}$ is the corresponding bimolecular collision rate ($\mathrm{m^{-3} \cdot s^{-1}}$).

The sticking coefficient is a dimensionless probability, between zero and one. In most versions of CNT, it is assumed that α is independent of cluster size and equals unity. Here we will make this assumption, meaning that we will assume that the forward reaction rate $R_{f,n}$ and the collision rate $Z_{1,n-1}$ are identical.

Assuming that monomers and all clusters behave as ideal gases, that molecules are hard spheres that interact only at the moment of collision, and that the gas is in translational equilibrium, the bimolecular collision rate obtained from ideal gas kinetic theory can be written as follows (Benson 1976; Vincenti and Kruger 1965):

$$Z_{AB} = \frac{N_A N_B}{S} \Omega_{AB} \overline{C_{AB}}, \tag{3.37}$$

where Z_{AB} is the rate per unit volume of all collisions between the molecules of species A and B, Ω_{AB} is the cross section ($\mathrm{m^2}$) for collisions between A and B molecules, $\overline{C_{AB}}$ is the mean thermal speed of molecules whose mass equals the reduced mass of the A-B collision pair, and the factor ζ in equation (3.37) equals two for the case of collisions between identical molecules, so as to avoid double-counting collisions for which the collision partners are indistinguishable, and unity otherwise.

Treating molecules and clusters as hard spheres, the collision cross section is given by

$$\Omega_{AB} = \pi d_{AB}^2, \tag{3.38}$$

where d_{AB}, the diameter of the collision cross section, is the average of the molecular or cluster diameters d_A and d_B, each derived assuming that clusters are spherical and have the same mass density as the bulk condensed phase.

The mean thermal speed of a molecule or cluster of mass m is given by

$$\overline{C} = \sqrt{\frac{8k_B T}{\pi m}}, \tag{3.39}$$

and the reduced mass m_{AB} of the A-B collision pair is defined by

$$m_{AB} \equiv \frac{m_A m_B}{m_A + m_B}. \tag{3.40}$$

More accurate treatments (see Section 6.5.3) account for the fact that collision rates are affected by intermolecular forces, resulting in a collision cross section diameter that differs from the simple hard-sphere model.

The mass of an n-mer is given by

$$m_n = n m_1. \tag{3.41}$$

Therefore, for the case of collisions between monomers and n-mers, the reduced mass from equation (3.40) is given by

$$m_{1,n} = \left(\frac{n}{n+1}\right) m_1. \tag{3.42}$$

Likewise, assuming that n-mers are spheres, one can write the diameter of the collision cross section as

$$d_{1,n} = r_1\left(1 + n^{1/3}\right), \tag{3.43}$$

where r_1 is the monomer radius.

Substituting equations (3.38) and (3.39) into equation (3.37) yields

$$Z_{AB} = \frac{N_A N_B}{s}\left(\sqrt{\frac{8\pi k_B T}{m_{AB}}} d_{AB}^2\right). \tag{3.44}$$

The term in parentheses in this equation is defined as the collision frequency function β_{AB} ($m^3 \cdot s^{-1}$). Substituting equations (3.42) and (3.43) into equation (3.44), and using $s_1 = 4\pi r_1^2$, one obtains

$$\beta_{1,n} = \sqrt{\frac{k_B T}{2\pi m_1}} s_1 \left(\frac{n+1}{n}\right)^{1/2} \left(n^{1/3} + 1\right)^2 \tag{3.45}$$

for the collision frequency function for collisions between monomers and n-mers. Combining equations (3.36), (3.44), and (3.45), and noting that $\zeta = 2$ for the case

$n = 2$ (dimer formation from monomer–monomer collisions) and unity otherwise, one finds

$$k_2 = 2\sqrt{\frac{k_B T}{\pi m_1}} s_1, \tag{3.46}$$

and

$$k_n = \sqrt{\frac{k_B T}{2\pi m_1}} s_1 \left(\frac{n}{n-1}\right)^{1/2} \left[(n-1)^{1/3} + 1\right]^2, n > 2, \tag{3.47}$$

for the condensation rate constant from hard sphere gas kinetic theory, neglecting intermolecular forces between the collision partners or nonunity sticking coefficients.

However, this is not the expression that is used in CNT. Rather, CNT uses the expression for the collision frequency function in the limit of very large clusters, $n \gg 1$. From equation (3.45),

$$\left(\beta_{1,n}\right)_{n \gg 1} = \lim_{n \to \infty} \beta_{1,n} = \sqrt{\frac{k_B T}{2\pi m_1}} s_1 n^{2/3} = \sqrt{\frac{k_B T}{2\pi m_1}} s_n. \tag{3.48}$$

As a cluster becomes very large, its mean thermal speed $\overline{C}$ approaches zero, and monomer–cluster collisions become like deposition of monomers on a flat surface. This result can therefore be compared to the expression from the kinetic theory of gases for the molecular flux β (m$^{-2}\cdot$s^{-1}) to a plane for a gas at translational equilibrium, where the dimensions of the plane are much larger than the molecular diameter,

$$\beta = \frac{N_1 \overline{C}}{4}. \tag{3.49}$$

Using equation (3.39) for $\overline{C}$, and the ideal gas law, gives

$$\beta = \sqrt{\frac{k_B T}{2\pi m_1}} N_1 = \frac{p_1}{\sqrt{2\pi m_1 k_B T}}. \tag{3.50}$$

Comparing equations (3.48) and (3.50), one can then write

$$\beta = \frac{N_1}{s_n} \left(\beta_{1,n}\right)_{n \gg 1}. \tag{3.51}$$

Classical nucleation theory makes the approximation that this expression for the monomer flux to an infinite flat plane can be used as the monomer flux to clusters of *all* sizes, regardless of how small. Thus the condensation rate is written as

$$R_n = \beta s_{n-1} N_{n-1}, \ n \geq 2. \tag{3.52}$$

In summary, comparing these alternate approaches, and assuming that the monomer sticking coefficient equals unity for clusters of all sizes, one can write for the condensation rate constant,

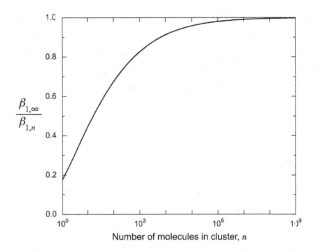

Figure 3.5 Ratio of the collision frequency function for monomer collisions with a stationary flat plane, as assumed by CNT, to the exact value that accounts for the cluster's finite size and motion.

$$k_n = \frac{\beta_{1,n-1}}{s} \approx \frac{\beta s_{n-1}}{N_1}, \ n \geq 2, \tag{3.53}$$

where the last approximation treats a cluster surface as being very large. Under this approximation, substituting equation (3.50) for β and equation (3.26) for the cluster surface area, one obtains

$$k_n = \sqrt{\frac{k_B T}{2\pi m_1}} s_1 (n-1)^{2/3}, \ n \geq 2. \tag{3.54}$$

This is the forward rate constant assumed in CNT.

Comparing equations (3.47) and (3.54), the ratio of the forward rate constant assumed in CNT to the "exact" kinetic theory result is given by

$$\frac{(k_n)_{\text{CNT}}}{(k_n)_{\text{exact}}} = \frac{\varsigma (n-1)^{2/3}}{\left(\frac{n}{n-1}\right)^{1/2} \left[(n-1)^{1/3} + 1 \right]^2}, \ n \geq 2, \tag{3.55}$$

where $\zeta = 2$ for $n = 2$, and unity otherwise.

Figure 3.5 compares the value of $\beta_{1,\infty}$ that is implicit in CNT, equation (3.48), to the exact value of the collision frequency function $\beta_{1,n}$, given by equation (3.45).[5] This approximation greatly simplifies the algebra in CNT, although, as can be seen, it is remarkably inaccurate until cluster sizes reach values far beyond those that are usually of interest for homogeneous nucleation. For $n < 10$, the CNT approximation

---

[5] The reason that Figure 3.5 plots the ratio of the collision frequency functions rather than the ratio of the rate constants given by equation (3.55) is simply to avoid the ζ term that applies in equation (3.55) at $n = 2$.

underpredicts the condensation rate by more than a factor of two, and even for $n = 100$, the approximation is low by about 33%. The approximation therefore introduces significant error into the final result for the nucleation rate, though it can be argued that it is acceptable because the error inherent in the classical model for the free energy of cluster formation is far more significant, often by orders of magnitude, as discussed in Section 3.12.4.

3.6.2   Evaporation Coefficient

Given the rate constant for condensation, the rate constant for evaporation from a cluster follows directly from equation (2.23):

$$k_{-n} = \frac{N^0 k_n}{K_{n-1,n}} = N^0 k_n \exp\left(\frac{\Delta G_{n-1,n}^0}{k_B T}\right), \ n \geq 2. \tag{3.56}$$

Note that while the standard reference pressure p^0 denoted by superscript "0" conventionally equals 1 atm or 1 bar, the choice of p^0 is arbitrary. From equation (2.76), the value of k_{-n} in equation (3.56) remains unchanged regardless of the choice of reference pressure as long as a consistent value is chosen for both the N^0 and ΔG^0 terms. For example, the standard reference pressure in equation (3.56) could be set to the equilibrium vapor pressure p_s. Thus one could rewrite equation (3.56) as

$$k_{-n} = N_s k_n \exp\left[\frac{\Delta G_{n-1,n}(p_s)}{k_B T}\right], \ n \geq 2, \tag{3.57}$$

where $N_s = p_s / k_B T$ is the equilibrium monomer number density at temperature T. Evaluating $\Delta G_{n-1,n}$ at $p = p_s$ from equation (3.23), and nondimensionalizing, one obtains

$$\frac{\Delta G_{n-1,n}(p_s)}{k_B T} = \left[n^{2/3} - (n-1)^{2/3}\right]\Theta, \ n \geq 2. \tag{3.58}$$

Substituting this result together with equation (3.54) into equation (3.57) yields

$$k_{-n} = \sqrt{\frac{k_B T}{2\pi m_1}} s_{n-1} N_s \exp\left\{\left[n^{2/3} - (n-1)^{2/3}\right]\Theta\right\}, \ n \geq 2. \tag{3.59}$$

for the evaporation rate constant, also known as the "evaporation coefficient."

It is worth noting that to close approximation the term in straight brackets in the argument of the exponential in equation (3.59) is given by

$$n^{2/3} - (n-1)^{2/3} \approx \frac{2}{3} n^{-1/3}, \tag{3.60}$$

as seen in Figure 3.6. The error in this approximation is quite small for all sizes larger than the dimer.

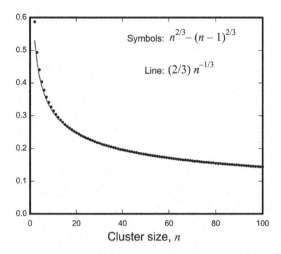

Figure 3.6 Function in argument of exponential in evaporation coefficient, equation 3.59.

3.7 Physical Properties in CNT

Having determined the predictions of classical theory for the condensation rate constant, k_n, equation (3.54), and the free energy of cluster formation, ΔG_n, equation (3.25), we are now in a position to use these expressions in equation (2.80) for the steady-state nucleation rate.

Before doing that, it is worth commenting on three substance-dependent physical properties that are required by the CNT expressions for rate constants and free energies: surface tension, mass density, and equilibrium vapor pressure.

3.7.1 Surface Tension

Classical nucleation theory assumes the capillarity model, in which surface tension is independent of cluster size and is equated to the value for an infinite flat surface of the liquid in equilibrium with its vapor. This is the value of surface tension that can be found for many liquids in a variety of reference sources. Even within the capillarity model, however, surface tension can vary considerably with temperature. For example, Figure 3.7 shows the surface tension of water in the temperature range 280–650 K, using values given at 10-K increments in the online NIST Chemistry Webbook (Lemmon et al. 2023).

A number of studies have attempted to modify CNT by considering the effect of surface curvature on surface tension, which then becomes size dependent. These approaches can be said to retain the general approach of classical theory, as the concept of surface tension is applied even down to very small cluster sizes, while avoiding the capillarity approximation.

In 1949, Tolman proposed a model for this effect (Tolman 1949), in which the size-dependent surface tension of a spherical droplet of radius r can be written as

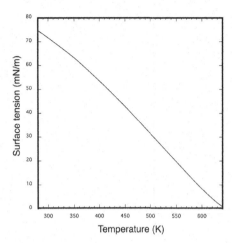

Figure 3.7 Surface tension of water versus temperature, from data in the online NIST Chemistry Webbook (Lemmon et al. 2023).

$$\frac{\sigma(r)}{\sigma_0} = \frac{1}{1 + 2\delta/r},$$
(3.61)

where σ_0 is the surface tension of a flat liquid surface of the same substance, in equilibrium with its vapor, and δ, known as the "Tolman length," is the thickness of a hypothetical layer, on the order of the molecular diameter, that separates the "surface of tension" from the "equimolar dividing surface."

The existence of surface tension causes a pressure difference between the interior of a liquid droplet and the surrounding gas. The work against surface tension is provided by this pressure difference. This can be written as

$$\left(p_l - p_g\right)dV = \sigma ds,$$
(3.62)

where p_l and p_g are, respectively, the pressures in the liquid droplet and the surrounding gas. Using, for a spherical droplet,

$$dV = 4\pi r^2 dr$$
(3.63)

and

$$ds = 8\pi r dr,$$
(3.64)

one obtains

$$p_l = p_g + \frac{2\sigma}{r}.$$
(3.65)

This relation is known as the Laplace equation. The "surface of tension" is the surface of a sphere whose radius satisfies this equation, where σ here is the planar surface tension of the capillarity approximation, denoted σ_0 in equation (3.61).

The molecular number density in a liquid droplet is typically much higher than in the surrounding vapor. In reality, the two phases may be separated by an interfacial layer of nonzero thickness, across which the number density makes a continuous transition between the two densities. However, in CNT the droplet surface is assumed to be sharply defined and to have zero thickness. Thus the number density is assumed to follow a step profile, dropping instantaneously from the liquid to the gas number densities. All of the liquid molecules are assumed to lie on the liquid side of this step, and the droplet is assumed to have uniform density. The location of this step defines the "equimolar surface."

In CNT, the surface of tension and the equimolar surface are identical. In Tolman's theory, they are separated by the distance δ that appears in equation (3.61). The value of δ for a number of substances at specific temperatures has been studied theoretically, and attempts have been made to modify CNT predictions of the critical cluster size and steady-state nucleation rate by incorporating Tolman's theory (e.g., Onischuk et al. 2006; Rao and McMurry 1990). Nevertheless, the value of the Tolman length, including its sign, remains the subject of considerable debate (Blokhuis and Kuipers 2006; Lei et al. 2005).

As the Tolman length is of the order of the intermolecular spacing in the liquid, it is evident from equation (3.61) that this effect becomes increasingly important as cluster size decreases, especially as it becomes only a small multiple times the molecular diameter. Unfortunately, this is precisely the small-cluster regime of relevance to nucleation, where the validity of macroscopic concepts such as surface tension becomes questionable. While the Tolman length has itself been estimated by atomistic approaches such as molecular dynamics simulations of fluids that obey a Lennard-Jones potential (Haye and Bruin 1994; Nijmeijer et al. 1992), the theory retains the macroscopic concept of a spherical liquid drop with surface tension and thus can be regarded as an attempt to patch classical theory without abandoning it.

3.7.2   Mass Density

The rate constants for condensation reactions depend directly on the corresponding collision cross sections. The free energy of cluster formation in CNT depends on the cluster surface area. The assumption in CNT that a cluster is spherical and has the same mass density as the bulk condensed phase provides a straightforward relation between the number of molecules in a cluster and its diameter (and collision cross section, assuming hard-sphere collision theory) and surface area. These assumptions are consistent with the general spirit of CNT, but are obviously crude approximations for clusters containing only a few or several tens of molecules.

3.7.3   Equilibrium Vapor Pressure

Measurements and predictions of steady-state nucleation rates typically characterize the nucleation rate of a substance as a function of temperature and saturation ratio. In CNT, there are several sources of temperature dependence. In addition to the $k_B T$ term in the denominator of the exponential in equation (2.80), Gibbs free energy and rate constants depend on temperature. As noted in Section 3.7.1 and illustrated in Figure 3.7, even under the capillarity approximation, surface tension is strongly temperature dependent.

An additional source of temperature dependence is revealed, though implicitly, by equation (2.81): The equilibrium vapor pressure, p_s, depends strongly on temperature, as given by the Clausius–Clapeyron relation. Assuming that the vapor behaves as an ideal gas and that the specific volume of the saturated liquid is much less than that of the saturated vapor at the same temperature, this relation can be written as

$$\frac{d\ln p_s}{dT} = \frac{L}{RT^2}, \tag{3.66}$$

where L is the molar enthalpy of vaporization (or "latent heat"), that is, the difference between the molar enthalpies of the coexisting vapor and liquid phases. Rearranging this expression gives

$$d\ln p_s = \frac{L}{RT^2}\,dT. \tag{3.67}$$

Treating L as approximately constant, and integrating, one obtains

$$p_s = C\exp\left(-\frac{L}{RT}\right), \tag{3.68}$$

where C is a constant that can be evaluated at any temperature T where $p_s(T)$ is known. Thus, for a given vapor partial pressure, the saturation ratio is exponentially temperature dependent.

3.8   The Steady-State Nucleation Rate

As given by equation (2.81), the steady-state nucleation rate can be written in terms of the free energy of cluster formation evaluated at the substance's equilibrium vapor pressure. In general, the forward rate constants in equation (2.81) may be the actual rate constants for chemical reactions involved in monomer addition or may incorporate three-body stabilization effects, etc. In CNT, it is assumed that equation (3.53) can be used, that is, that the condensation rate constants can be replaced by a term that approximates the monomer flux to a cluster as being the same as that to a plane, and that assumes a sticking coefficient of unity. Then, using equation (3.53) for k_n, and using the fact that for spherical droplets the surface area $s_n \propto n^{2/3}$, one can write

$$k_{n+1} = \frac{\beta s_1 n^{2/3}}{N_1}. \tag{3.69}$$

If in addition one uses equation (3.29) for the CNT model for the Gibbs free energy of cluster formation evaluated at the equilibrium vapor pressure p_s, equation (2.81) can be rewritten as

$$J = \frac{\beta s_1 N_1}{\displaystyle\sum_{n=1}^{M}\left\{n^{2/3}S^{n-1}\exp[-(n^{2/3}-1)\Theta]\right\}^{-1}}. \tag{3.70}$$

The summand is now dimensionless. Let us define a dimensionless function $H(n)$ such that equation (3.70) can be rewritten as

$$J = \frac{\beta s_1 N_1}{\displaystyle\sum_{n=1}^{M} e^{-H(n)}},$$

(3.71)

where by inspection,

$$H(n) = \ln\left(n^{2/3} S^{n-1}\right) - \left(n^{2/3} - 1\right)\Theta.$$

(3.72)

Now, as M is arbitrarily large, let us make the approximation that the summation over $n = 1$ to M can be replaced by an integral over $x = 0$ to ∞, with the discrete cluster size variable n replaced by the continuous size variable x:

$$\sum_{n=1}^{M} e^{-H(n)} = \sum_{n=1}^{M} e^{-H(n)} \Delta n \approx \int_{0}^{\infty} e^{-H(x)} dx,$$

(3.73)

where the increment Δn between adjacent sizes equals unity.

To evaluate the integral, let us assume that there exists some size x' at which the first derivative of $H(x)$ equals zero. Then, expanding $H(x)$ in a Taylor series about that size, we have

$$H(x) \approx H(x') + (x-x')\left(\frac{dH}{dx}\right)_{x=x'} + \frac{1}{2}(x-x')^2\left(\frac{d^2H}{dx^2}\right)_{x=x'} + \cdots,$$

(3.74)

where the second term on the right-hand side equals zero. Thus, neglecting terms of order higher than the second derivative term, we have

$$\int_{0}^{\infty} e^{-H(x)} dx = e^{-H(x')} \int_{0}^{\infty} \exp\left[-\frac{H''(x')}{2}(x-x')^2\right] dx,$$

(3.75)

where $H''(x')$ stands for the second derivative of $H(x)$, evaluated at $x = x'$. The integrand on the right-hand side of equation (3.75) can be simplified by defining

$$a \equiv \frac{H''(x')}{2}$$

(3.76)

and

$$y \equiv x - x'.$$

(3.77)

Here a is simply a constant – whatever the value of H'' happens to equal at $x = x'$, divided by two. With these substitutions, we have

$$\int_{0}^{\infty} e^{-H(x)} dx = e^{-H(x')} \int_{-x'}^{\infty} e^{-ay^2} dy.$$

(3.78)

The integral is seen to be close in form to the error function,

$$\mathrm{erf}\, z \equiv \frac{2}{\sqrt{\pi}} \int_{0}^{z} e^{-u^2} du.$$

(3.79)

Noting that $\mathrm{erf}\, z \to 1$ as $z \to \infty$ and that the integrand in equation (3.79) is symmetric about $y = 0$, the value of the integral on the right-hand side of equation (3.78) is found to be given by

$$\int_{-x'}^{\infty} e^{-ay^2} dy = \frac{1}{2} \left(\frac{\pi}{a} \right)^{1/2} \left[1 + \mathrm{erf}\left(a^{1/2} x' \right) \right]. \tag{3.80}$$

Utilizing this intermediate result would cause the final result for the nucleation rate J to contain an error function. However, the value of the integral can be cast into a simpler form if we make the approximation that the lower limit of integration $-x'$ in equation (3.80) is replaced by $-\infty$, producing an integral whose value is

$$\int_{-\infty}^{\infty} e^{-ay^2} dy = \sqrt{\frac{\pi}{a}}. \tag{3.81}$$

Classical nucleation theory makes this approximation. As the integrand here is symmetric about $y = 0$, the effect of replacing the lower limit of integration $-x'$ with $-\infty$ results in an overestimation of the value of the integral by at most a factor of two. Because the summand that the integral replaces appears in the denominator of equation (3.71), the final effect of this approximation is to underestimate the nucleation rate, again by a factor of at most two. As discussed in connection with the "very large cluster" approximation for monomer–cluster collisions, factors that produce errors of less than an order of magnitude are relatively inconsequential in nucleation theory. Again this is because uncertainties in the Gibbs free energies of cluster formation, which appear in the argument of the exponential of equation (2.81), usually produce far greater errors in the final result. However, this "lower limit of the integral" approximation produces an error that is in the same direction as the "stationary cluster" approximation – underpredicting the nucleation rate – so that the total error incurred in making these two approximations is not necessarily negligible.

In any case, from equations (3.76), (3.78), and (3.81), we then have

$$\int_{0}^{\infty} e^{-H(x)} dx \approx e^{-H(x')} \sqrt{\frac{2\pi}{H''(x')}}. \tag{3.82}$$

Replacing the summation in equation (3.71) by this result gives

$$J = \beta s_1 N_1 \sqrt{\frac{H''(x')}{2\pi}} \exp[H(x')]. \tag{3.83}$$

The value of $H(x)$ and its second derivative must now be evaluated at a location $x = x'$ where its first derivative equals zero. From equation (3.72), with the integer size variable n converted to the continuous size variable x, $H(x)$ is given by

$$H(x) = \frac{2}{3} \ln x + (x - 1) \ln S - \left(x^{2/3} - 1 \right) \Theta, \tag{3.84}$$

so that

$$\frac{dH}{dx} = \frac{2}{3} x^{-1} + \ln S - \frac{2}{3} \Theta x^{-1/3}. \tag{3.85}$$

Setting this last expression to zero and rearranging, one finds that x' is the solution of

$$\frac{\Theta(x')^{2/3} - 1}{x'} = \frac{3}{2} \ln S. \tag{3.86}$$

As noted previously, values of Θ for most realistic scenarios range from about 3 to 50. The value of x' must be greater than unity, and indeed the validity of the liquid droplet model implicitly rests on the assumption that $x' \gg 1$. Therefore, in the numerator of the left-hand side of equation (3.86), let us make the approximation that $\Theta(x')^{2/3} \gg 1$. Equation (3.86) then reduces to

$$x' = \left(\frac{2}{3} \frac{\Theta}{\ln S} \right)^3. \tag{3.87}$$

Comparing this result to equation (3.32), we see that the value of x' is identical to that of the critical cluster size x^*. However, the derivation of the critical cluster size x^* was based on physical reasoning – the size for which the Gibbs free energy of cluster formation is a maximum – whereas the above result for x' followed from a series of arguments that were made purely for mathematical convenience, beginning with the need to evaluate the integral in equation (3.73). Indeed, if it were not for the approximation $\Theta(x')^{2/3} \gg 1$ in equation (3.86), x' and x^* would differ.

Differentiating equation (3.85), the second derivative $H''(x)$ is given by

$$H''(x) = -\frac{2}{3}x^{-2} + \frac{2}{9}\Theta x^{-4/3}. \tag{3.88}$$

Evaluating both H and H'' at $x = x'$, where x' is given by equation (3.87), and inserting these results into equation (3.83) yields

$$J = \frac{\beta s_1 N_1}{3S} \sqrt{\frac{\Theta}{\pi}} \exp\left[\Theta - \frac{4}{27} \frac{\Theta^3}{(\ln S)^2} \right]. \tag{3.89}$$

Then, substituting for β from equation (3.50), and using the relationship between the diameter and surface area of the (assumed spherical) monomer, we finally obtain

$$\tag{3.90}$$

$$\boxed{J = \frac{d_1^2}{6} \sqrt{\frac{2k_B T \Theta}{m_1}} [N_s(T)]^2 S \exp\left[\Theta - \frac{4}{27} \frac{\Theta^3}{(\ln S)^2} \right].}$$

This expression gives the final result for the steady-state rate of single-component homogeneous nucleation, in the self-consistent classical theory.

3.9   Relation of Nucleation Rate to Properties of the Critical Cluster

While equation (3.90) provides an explicit recipe for the nucleation rate as a function of saturation ratio, temperature, and substance properties, various rearrangements are possible that provide more direct physical insight. In particular, returning to equation

(3.89), one finds that it can be recast in various ways that emphasize the relation between the nucleation rate and properties of the critical cluster.

3.9.1   The Energy Barrier to Nucleation

Comparing equations (3.27), (3.32), and (3.89), one finds that equation (3.90) can be rewritten as

$$J = J_0 \exp\left[-\frac{\Delta G^*(p_1)}{k_B T}\right], \tag{3.91}$$

where

$$J_0 = \frac{\beta s_1 N_1}{3} \sqrt{\frac{\Theta}{\pi}} \tag{3.92}$$

and

$$\frac{\Delta G^*(p_1)}{k_B T} = \frac{4}{27} \frac{\Theta^3}{(\ln S)^2} - \Theta + \ln S. \tag{3.93}$$

Equation (3.91) makes clear that the free energy of formation of the critical cluster constitutes the energy barrier to nucleation and is analogous to the activation energy of a chemical reaction. As given by equation (3.93), the energy barrier is a function of saturation ratio, surface tension, and temperature.

We further note that the term $\beta s_1 N_1$ in equation (3.92) is equal, in CNT, to the forward rate of dimerization, R_2. If the energy barrier to nucleation were zero – corresponding to the collision-controlled regime discussed in Chapter 1 and to the lowest curve in Figure 3.2 – then the nucleation rate would be identical to the dimerization rate. Indeed the factor $(1/3)\sqrt{\Theta/\pi}$ in equation (3.92), with Θ lying in the range $\sim$ 3–50 for most liquids and realistic scenarios, is of order unity. The fact that it does not precisely equal unity arises due to various approximations made in CNT, notably including the treatment of cluster size as a continuous function, whereas the concept of "dimerization" implicitly treats cluster size as a discrete quantity.

3.9.2   Nucleation and the Equilibrium Number Density of Critical Clusters

Another form of equation (3.90) that is often found in the literature can be written as

$$J = Zf^* \times (N_{n*})_{eq}, \tag{3.94}$$

where f^* is the rate (s^{-1}) of monomer addition to a single critical cluster, $(N_{n*})_{eq}$ represents the number density of critical clusters that would exist if they were in equilibrium with the actual monomer number density N_1, and Z is a dimensionless term known as the "Zeldovich factor."

The term $(N_{n*})_{eq}$ is related to the equilibrium constant for the reaction $n^* A_1 \rightarrow A_{n*}$ – that is, to the exponential term in equation (3.91) – by

$$\exp\left[-\frac{\Delta G^*(p_1)}{k_B T}\right] = \left[\frac{p_{n*}/p_1}{(p_1/p_1)^{n*}}\right]_{eq} = \left(\frac{N_{n*}}{N_1}\right)_{eq} = \frac{(N_{n*})_{eq}}{N_1}, \qquad (3.95)$$

where the standard reference pressure p^0 has been set equal to p_1.

In CNT, the rate of monomer addition to a single critical cluster, assuming unity sticking coefficient, is given by

$$f^* = \beta s_{n*}. \qquad (3.96)$$

Thus the product $f^* \times (N_{n*})_{eq}$ gives what the rate (m^{-3}·s^{-1}) of monomer addition to critical clusters would equal if the number density of critical clusters were in equilibrium with the number density of monomers.

Now, as monomer addition to clusters at and above the critical size can be treated as effectively irreversible, the nucleation rate can be equated with the rate of the reaction in which a monomer condenses onto a critical cluster,

$$A_{n*} + A_1 \rightarrow A_{n*+1}. \qquad (3.97)$$

Thus the factor Z in equation (3.94) can be interpreted as a nonequilibrium correction, which, when multiplied by $(N_{n*})_{eq}$, gives the *actual* number density of critical clusters during steady-state nucleation. Considering equations (3.91), (3.92), (3.94), and (3.95); using equation (3.32) for n^*; and noting from equation (3.26) that

$$s^* = (x^*)^{2/3} s_1 = \left(\frac{2}{3}\frac{\Theta}{\ln S}\right)^2 s_1, \qquad (3.98)$$

where s^* is the surface area of the critical cluster, one finally obtains

$$Z = \frac{3(\ln S)^2}{4\pi^{1/2}\Theta^{3/2}} \qquad (3.99)$$

for the value of the Zeldovich factor.

Within the condensation/evaporation regime, where $1 < S < S^*$, one finds that $Z < 1$,[6] implying that during steady-state nucleation the actual number density of critical clusters is lower than the number density that would exist in equilibrium with a given number density of monomers. Equivalently, the number density of monomers is in superequilibrium with respect to the actual number density of critical clusters. Indeed that must obviously be the case, consistent with there being a driving force for nucleation. For example, for $\Theta = 10$ and $S = 10$, one finds $Z \approx 0.071$. For most substances and conditions of interest, Z lies in the range $\sim 10^{-2}$ to 1.

Equation (3.94) is often used in the literature as a straightforward way to calculate steady-state nucleation rates predicted by CNT, using as inputs the free energy of formation and surface area of the critical cluster, with the value of n^* calculated using equation (3.32).

---

[6] For values of S close to S^* as given by equation (3.30), one finds that Z may slightly exceed unity. Again this is due to the fact that equation (3.30) treats cluster size as a discrete function, whereas our derivation of equation (3.99) is based on the assumption that cluster size is a continuous function.

This approach is appealing, as it represents an intuitively reasonable way of representing the nucleation rate. Moreover, it is tempting to argue that equation (3.94) provides support for the classical model, as it appears to require knowledge of properties only of the critical cluster not of smaller clusters, and in some (though certainly not all) cases the critical cluster may be large enough for the classical model to seem a reasonable approximation.

However, this argument is misleading, for several reasons. Of these, the most important concern is the value of ΔG^* and the value of n^* itself. These issues are discussed in Sections 3.11.6, 3.12.4, and Chapter 6.

3.9.3   Classical Theory and the Nucleation Theorem

From equation (3.90), for a given substance at fixed temperature, one can write

$$J = C_1 S \exp\left[-\frac{4}{27} \frac{\Theta^3}{(\ln S)^2} \right],$$
(3.100)

where C_1 is a constant. Thus

$$\ln J = C_2 + \ln S - \frac{4}{27} \frac{\Theta^3}{(\ln S)^2},$$
(3.101)

where C_2 is some other constant. Differentiating this expression with respect to $\ln S$, with Θ held constant as the temperature is fixed, gives

$$\left(\frac{\partial \ln J}{\partial \ln S} \right)_T = 1 + \frac{8}{27} \frac{\Theta^3}{(\ln S)^3}.$$
(3.102)

Comparing this result to the expression for the size of the critical cluster, equation (3.32), one finally obtains

$$\left(\frac{\partial \ln J}{\partial \ln S} \right)_T = n^* + 1.$$
(3.103)

Thus the nucleation theorem is exactly correct in CNT (at least in the self-consistent version; see also Section 3.10).

3.10   The "Standard Version" of Classical Theory

As noted in Section 3.3.4, the "standard version" of CNT writes the Gibbs free energy of cluster formation as

$$\Delta G_n(p_1) = \sigma s_n - n k_B T \ln S,$$
(3.104)

rather than as in equation (3.25).

The lack of self-consistency in equation (3.104) lies in the fact that it yields, for the free energy of formation of a monomer,

$$\Delta G_1(p_1) = \sigma s_1 - k_B T \ln S, \tag{3.105}$$

which does not in general equal zero, whereas equation (3.25) yields $\Delta G_1 = 0$ for all values of S. As the free energy of formation of a monomer represents the change in free energy for the trivial reaction

$$A_1 \rightarrow A_1,$$

ΔG_1 must equal zero regardless of the model used to estimate cluster free energies, as in equation (2.30). A nonzero ΔG for the above reaction implies a nonunity equilibrium constant, meaning that

$$\left(\frac{p_1}{p_1}\right)_{eq} \neq 1. \tag{3.106}$$

As this is obviously incorrect, equation (3.105) may be interpreted as a violation of the law of mass action.[7] Furthermore, this error in ΔG_1 causes the values of ΔG_n for all values of n to be offset from their "correct" value, as given by equation (3.25).

Nevertheless, it is easy to see how equation (3.104) arises. Equation (3.91) implies that the only free energy of formation that matters in the expression for the steady-state nucleation rate is that of the critical cluster. If the critical cluster is sufficiently large – and implicitly the classical model assumes this – then the error in ΔG_n in writing equation (3.104) instead of equation (3.25) might be thought to be negligible, especially in comparison with other approximations that are made. However, because ΔG^* in equation (3.91) is exponentiated, the error in the final result for the nucleation rate is far from negligible. Inserting either equation (3.25) or equation (3.104) into equation (3.91), with the pre-exponential term J_0 being the same in either theory, one obtains

$$J_{\text{self-consistent}} = J_{\text{standard}} \times \frac{e^\Theta}{S}. \tag{3.107}$$

As Θ for must substances and realistic conditions lies in the range ~3–50, the resulting error can easily equal several orders of magnitude. Of course, it may still be argued that, if one is going to make use of the classical liquid droplet model, which cannot be expected to give accurate results in any case, then one may as well use equation (3.104), which seems more consistent with the simplicity of the classical model than does equation (3.25). Indeed, while the "self-consistency" correction presented here, as well as other proposed modifications, has been the subject of much debate, the standard version of the nucleation rate is what one still sees most often in the literature.

It should also be noted that the nucleation theorem is slightly different in the standard version of CNT than in the relation given by equation (3.103). Following the same

---

[7] It has been proposed that the self-consistency problem can be avoided by treating vapor monomers and liquid monomers as separate species (Reguera et al. 2003). In that case, the first step in the reaction sequence would be $A_1(v) \rightarrow A_1(l)$, which indeed would have a nonzero value of ΔG. However, this approach would involve assigning separate number densities to "vapor monomers" and "liquid monomers," even though it is physically impossible to distinguish between them. And if it were possible, then the rate of this reaction would have an important effect on the nucleation rate, but it is not evident how one would assign a rate constant.

derivation as in Section 3.9.3, but with the nucleation rate now modified by equation (3.107), one obtains

$$\left(\frac{\partial \ln J}{\partial \ln S}\right)_T = n^* + 2 \tag{3.108}$$

for the version of the nucleation theorem that corresponds with the standard version of CNT.

The discrepancy between equations (3.103) and (3.108) can be traced to the $nk_BT \ln S$ term in equation (3.104), which should instead be written as $(n-1)k_BT \ln S$, as in equation (3.25). Given this discrepancy, together with the fact that the experimental determination of the slope of $\ln J$ versus $\ln S$ along an isotherm can involve considerable uncertainty, what one often finds in the literature is the approximation,

$$\left(\frac{\partial \ln J}{\partial \ln S}\right)_T \approx n^* \tag{3.109}$$

(Bhabhe and Wyslouzil 2011; Manka et al. 2010, 2012; Sinha et al. 2010).

3.11   Comparisons to Experimental Data

Beginning with Wilson's cloud chamber, a large number of experimental studies have been conducted to study gas-phase nucleation of a variety of substances. Up until about 1980, these experiments measured the "critical supersaturation" (occasionally termed the "onset supersaturation"), that is, the minimum saturation ratio required for observable nucleation to occur. Subsequent improvements in experimental techniques allowed measurements of nucleation rates themselves.

3.11.1   Experimental Methods

Experimental methods for studying gas-phase nucleation all involve bringing a condensible vapor from a state where no observable nucleation occurs to a state of abundant nucleation. Because nucleation rate is a strong function of saturation ratio, typically this transition is sharp. Most experimental methods are based either on the diffusion of a vapor through a background gas from hotter toward colder regions or on a temperature drop caused by expansion. If the partial pressure of the vapor is fixed, then as temperature drops the equilibrium vapor pressure drops, as given by equation (3.68), and therefore the saturation ratio increases. In many of the methods, the vapor partial pressure itself decreases in the direction of falling temperature, but the equilibrium vapor pressure, being exponentially temperature dependent, decreases even faster, again causing the saturation ratio to increase.

Starting with monomers of a condensible vapor, finite time is required for a population of subcritical clusters to form and for a steady-state nucleation current to become established, as discussed in Chapter 7. Therefore, as temperature drops, the saturation ratio may rise to a high value before observable nucleation occurs. Higher cooling rates typically cause the saturation ratio to overshoot unity by a greater amount. Nucleation itself, as well as surface growth by vapor condensation on nucleated particles, causes the vapor to be depleted, and, depending on conditions, the temperature may rise due to the heat release accompanying condensation. Both of these effects tend to cause the saturation ratio to relax toward unity, quenching nucleation. Hence, nucleation often occurs in a rapid burst, in which the nucleation rate rises sharply and then falls. Therefore methods that inherently involve lower cooling rates, such as diffusion across an externally imposed temperature gradient, are suitable for exploring nucleation at relatively low saturation ratios, while methods that involve rapid cooling, such as supersonic expansions, are able to probe nucleation at high saturation ratios.

A brief overview of experimental methods is provided here. For more detail, the reader is referred to the reviews by Heist and He (1994), Wyslouzil and Wölk (2016), and references therein.

Diffusion-Based Methods

In diffusion-based methods, condensible vapor diffuses through an inert carrier gas across a temperature gradient, with the vapor diffusing from hotter to colder regions, causing its saturation ratio to increase until nucleation occurs. Techniques include systems both with and without convective flow of the bulk gas, the former being termed laminar flow diffusion devices and the latter thermal diffusion cloud chambers.

In either case, measurements of conditions such as wall temperatures, total pressure, and flow rate (if there is bulk flow), together with the solution of the pertinent equations of mass, momentum, and energy conservation, allow the determination of detailed spatial profiles of temperature and vapor partial pressure. Knowledge of the equilibrium vapor pressure–temperature relationship of the condensible vapor then provides spatial profiles of the vapor saturation ratio.

Particle formation is typically observed by light scattering. The location where light scattering is first observed yields an estimate of the critical supersaturation. Various experimental methods have also been developed for counting the droplets formed per unit time, and hence the nucleation rate.

In an upward thermal diffusion cloud chamber, condensible vapor is generated by heating a pool of liquid maintained at a fixed temperature on top of a bottom plate. Evaporated vapor diffuses upward through an inert background gas, across a gap bounded by a cold top plate. The plates are surrounded by a transparent ring, whose diameter is much greater than the gap between the plates, so that the system can be treated with good accuracy as one dimensional, varying only in the direction normal to the two plates.

Typical profiles of total gas-phase mass density, temperature, partial pressure of the condensible vapor, saturation ratio, and nucleation rate are shown in Figure 3.8 (Brus et al. 2005). This method exploits the fact that equilibrium vapor pressure is an exponential function of temperature and that the nucleation rate is a strong function of saturation ratio. Thus, as vapor diffuses upward, its saturation ratio increases until a sharp burst of nucleation occurs. Nucleation and particle surface growth relieve the supersaturation, which then declines toward unity in regions lying above the nucleation burst.

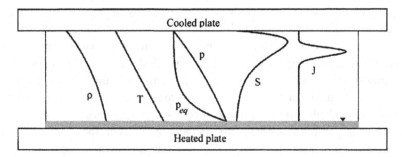

Figure 3.8 Typical profiles in a thermal diffusion cloud chamber of total gas-phase density, temperature, equilibrium (saturation) vapor pressure of the condensible vapor, actual partial pressure of the condensible vapor, saturation ratio, and nucleation rate. Reprinted from Brus et al. (2005), with the permission of AIP Publishing.

In laminar flow diffusion devices, a warm vapor-laden gas flows down a tube with cold walls. Because heat transfer to the walls is faster than vapor diffusion, the vapor becomes supersaturated as the gas cools in the flow direction, causing nucleation to occur.

Expansion Methods

Expansion methods rely on the fact that dropping the pressure or increasing the volume of a gas causes its temperature to drop, leading to supersaturation of a condensible vapor. From the Gibbs equation for a simple compressible substance,

$$TdS = dE + pdV, \qquad (3.110)$$

where S here is entropy, it follows that for an isentropic expansion,

$$\frac{dE}{T} = -\frac{p}{T}dV. \qquad (3.111)$$

Assuming ideal gas behavior and employing the definition of specific heat, this can be rewritten as

$$c_v \frac{dT}{T} = -R\frac{dV}{V} = -(c_p - c_v)\frac{dV}{V}, \qquad (3.112)$$

where c_v and c_p are the specific heats at constant volume and constant pressure, respectively. Thus

$$\frac{dT}{T} = -(\gamma - 1)\frac{dV}{V}, \qquad (3.113)$$

where γ is the ratio of specific heats, c_p/c_v. Integrating over a change in volume from V_1 to V_2, and assuming constant specific heats, one then obtains the corresponding temperature change as

$$\frac{T_2}{T_1} = \left(\frac{V_1}{V_2}\right)^{\gamma-1}. \qquad (3.114)$$

Alternatively, if one instead uses the enthalpy form of the Gibbs equation, equation (2.72), one obtains

$$\frac{T_2}{T_1} = \left(\frac{p_2}{p_1}\right)^{(\gamma-1)/\gamma}. \tag{3.115}$$

For any gas, γ is always greater than unity. Thus for any gas, an increase in volume or a decrease in pressure in an isentropic (adiabatic and reversible) process always causes a drop in temperature. Often the inert carrier gas is a noble gas, either argon or helium, for which constant specific heat is an excellent assumption over accessible temperature ranges, and $\gamma = 5/3$. Alternatively, a diatomic gas such as N_2, which is effectively inert over temperatures of interest, or H_2, under conditions where it is effectively inert, is often used, and $\gamma = 7/5$ at room temperature and below, decreasing somewhat at elevated temperature.

Expansion-type devices for nucleation studies include those based on movable pistons for increasing the gas volume, and shock tubes and supersonic nozzles for decreasing the gas pressure.

Wilson's late-nineteenth-century cloud chamber was an expansion device, involving a movable piston. A later development involved the introduction of a recompression stroke, by using either the same piston as for the initial expansion or a second piston. In this way, the nucleation burst could be rapidly terminated, sharply defining the nucleation pulse and limiting its duration to timescales on the order of milliseconds. This type of device is often referred to as a "nucleation pulse chamber."

In a shock tube, a high-pressure chamber containing the mixture of condensible vapor and inert carrier gas is separated by a diaphragm from a vacuum chamber. When the diaphragm is broken, the gas rapidly expands into the low-pressure region. The initial pressure difference between the two regions is large enough that a thin shock forms downstream of the broken diaphragm. Pressure and temperature drop steeply across the shock, causing nucleation to occur.

Whereas pistons and shock tubes are batch devices, supersonic nozzles can involve continuous flow or can be pulsed. Typically Laval nozzles are used, which involve a converging section followed by a diverging section, with the pressure ratio across the nozzle sufficiently high to drive supersonic flow downstream of the nozzle throat. Velocity and temperature profiles are controlled by the nozzle geometry and the ratio of the pressures upstream and downstream of the nozzle. Flow acceleration can be quite rapid, so that the temperature drops rapidly and the saturation ratio climbs to high values before observable nucleation occurs. Supersonic nozzles typically achieve the highest nucleation rates of any of the devices employed to study nucleation, and the nucleation burst, quenched both by vapor depletion and the temperature rise due to heat release accompanying rapid condensation, can be quite short, typically on the order of 10^{-5} s.

3.11.2   Critical Supersaturation

As changes of only a few percent in saturation ratio can produce order-of-magnitude changes in nucleation rate, the value of the critical supersaturation is a fairly robust characteristic of nucleation for a given substance at given temperature. However, from the theoretical viewpoint, any saturation ratio greater than unity results in a finite nucleation rate, but that rate may be far too small to be observable in experiments. Moreover, the value of the nucleation rate that qualifies as "observable" depends on the apparatus and measurement technique. Thus any comparison of experimental critical supersaturations with theoretical predictions requires one to specify a criterion for the nucleation rate that corresponds to the "critical supersaturation." The critical supersaturation has typically been defined as the value of the saturation ratio for which the nucleation rate equals some arbitrary value. For example, values of 1, 10^6, 10^7, and 10^{12} cm$^{-3}\cdot$s^{-1} are all found in the literature. Thus there is some degree of arbitrariness in the value of the nucleation rate at which the critical supersaturation is evaluated, and caution must be exercised in comparing different studies of the same substance. However, the difference in critical supersaturation for cases with nucleation rates that differ by many orders of magnitude, while not negligible, is often impressively modest.

Conversely, it follows that experimental determination of the critical supersaturation is a rather insensitive test of CNT's prediction of the nucleation rate. Nevertheless, the reasonably good agreement between experiment and CNT for the critical supersaturation for most substances tested can justifiably be regarded as a triumph of the theory. Oxtoby, in his 1992 review of nucleation experiments, observed that measured critical supersaturations agreed within 10% of the prediction of the standard version of CNT for the vast majority of substances that had been tested to that date (Oxtoby 1992). After that date, published studies tended to focus on measured nucleation rates – a much more sensitive test of the theory – rather than on critical supersaturation.

3.11.3 Measurements of Nucleation Rates

A large number of experimental studies have been reported in which homogeneous nucleation rates of various substances were directly measured as a function of temperature and vapor saturation ratio. Usually these measurements are compared in the literature with predicted nucleation rates based on the standard version of CNT and/or its various modifications and alternatives. Typically the experimental results are presented on log–log plots of nucleation rate versus saturation ratio, with data points organized along isotherms. As noted in Section 3.11.1, a given experimental system is capable of generating and measuring only a finite range of nucleation rates, corresponding to a particular region of temperature-saturation-ratio space, depending on the substance. For example, Figure 3.9 shows measured nucleation rates from a study of nucleation of light water, H_2O, versus its heavier deuterium isotope, D_2O (Wölk and Strey 2001). The experimental apparatus employed was a two-piston expansion chamber (or "nucleation pulse chamber"), which allowed measurements of nucleation rates ranging from about 10^5 to 10^9 cm^{-3} s^{-1}. Thus all the data in the figure lie in this range of nucleation rates. The horizontal dashed line in the figure, at a nucleation rate of 10^7 cm^{-3} s^{-1}, denotes the value that the authors used as the criterion for evaluating the critical supersaturation. Clearly, this choice was made because it lay roughly in the middle, on a logarithmic scale, of the range of nucleation rates accessible by their experimental apparatus.

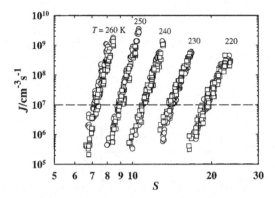

Figure 3.9 Measured nucleation rates of light water (circles) and heavy water (squares), versus saturation ratio, at several temperatures. Horizontal dashed line denotes the criterion the authors used to determine the critical supersaturation. Reprinted with permission from Wölk and Strey (2001). Copyright 2001 American Chemical Society.

As can be seen, the isotherms in Figure 3.9 are steep, implying that the dependence of nucleation rate on saturation ratio is strong. For a given saturation ratio, the nucleation rate is also seen to be a strong function of temperature, with each 10-degree increase in temperature resulting in an increase in the nucleation rate by roughly two orders of magnitude. Therefore, with the experimental apparatus accessing a given range of nucleation rates, the data for different isotherms can be arranged side by side on such a plot, with isotherms lying increasingly to the right as temperature decreases, implying that as temperature decreases a larger saturation ratio is required to achieve the same nucleation rate. Additionally, one notes that the isotherms in Figure 3.9 become progressively less steep as the temperature decreases. According to the nucleation theorem, equation (2.86), this implies that the critical cluster size is smaller in the lower temperature cases, again when comparing experiments whose nucleation rate is the same. Thus, as temperature decreases, achieving a given nucleation rate requires a higher saturation ratio, corresponding to a smaller critical cluster size.

These sorts of measurements have been made for many substances. For example, Figure 3.10 shows measurements of nucleation rates of a series of 1-alcohols ($C_nH_{2n+1}OH$, n = 2-5), using two different nucleation pulse chambers (Manka et al. 2012). Dashed lines in the figure show the corresponding predictions of the standard version of CNT at temperatures (right to left) of 235, 245, 255, and 265 K. Qualitatively, the comments made regarding the results for nucleation of water shown in Figure 3.9 apply as well to the 1-alcohols in Figure 3.10, although the quantitative results obviously differ quite substantially from substance to substance.

Thus, for both water and the 1-alcohols, as well as for many other substances studied, experimental results indicate that as temperature decreases achieving a given nucleation rate requires a higher saturation ratio. Indeed a straightforward explanation is available, which indicates that this conclusion may be somewhat misleading. To see why, consider the nucleation rate given by equation (2.68), which is independent of the model used to evaluate the Gibbs free energy of cluster formation. As discussed in Section 2.7, the N_1^2 term in front of the summation arises from purely kinetic considerations. But the

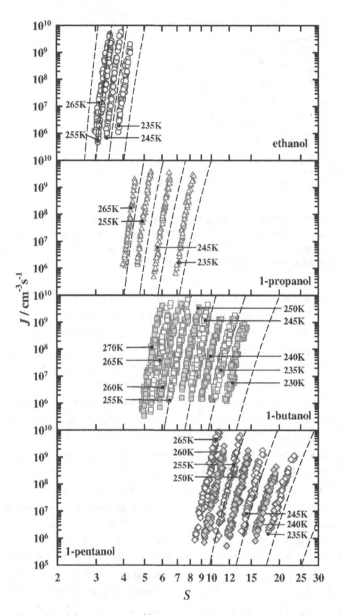

Figure 3.10 Measurements of homogeneous nucleation rates of the 1-alcohols ($C_2H_{2n+1}OH$, $n=2$–5). Data for ethanol, 1-propanol, and 1-butanol are from Manka et al. (2012), while the 1-pentanol measurements are from Iland et al. (2004). Dashed lines show predictions of standard version of CNT at 235, 245, 255, and 265 K. Reprinted from Manka et al. (2012) with the permission of AIP Publishing.

relationship between the nucleation rate and N_1^2 really has nothing to do with saturation ratio, even though $N_1 = SN_s(T)$ and thus can be replaced in favor of S, as done, for example, in going from equation (2.84) to (2.85). As temperature decreases, the saturation vapor pressure exponentially decreases, according to equation (3.68), and thus maintaining the same monomer number density at a given temperature requires a higher saturation ratio.

Likewise, as temperature increases a smaller saturation ratio is required to achieve the same monomer number density. This effect alone is sufficient to explain the qualitative observations that are evident in Figures 3.9 and 3.10, regarding the effect of temperature on the saturation ratio required to achieve a given nucleation rate.

Inspecting the experimental data in Figure 3.10, one observes that as one moves progressively from ethanol to 1-pentanol, that is, from lighter to heavier alcohols, the saturation ratio required to achieve a given nucleation rate increases. The dashed lines in the figure, showing the predictions of CNT, are qualitatively consistent with this trend, which in CNT must be attributable to differences in monomer mass and diameter, surface tension of the bulk liquid, and equilibrium vapor pressure, as those are the only substance properties that enter into the CNT prediction. Indeed, in general higher alcohols have lower equilibrium vapor pressures, which makes the purely kinetic effect of N_1^2 again the most straightforward explanation for this behavior.

3.11.4   Comparisons with Classical Theory: Magnitude of Nucleation Rate

The results for the 1-alcohols in Figure 3.10, together with results for the same substances from a number of other studies, are replotted in Figures 3.11 (for ethanol, 1-propanol, and 1-butanol) and 3.12 (for 1-pentanol) in terms of the ratio of the measured nucleation rate to the rate predicted by the standard version of CNT, as a function of inverse temperature. Each data point on these graphs represents an average for a given measurement series (i.e., isotherm) in a given experimental study. The results of similar exercises are presented in Figures 3.13–3.16, from studies of nucleation of n-nonane (C_9H_{20}), water[8], argon, and N_2, respectively. Measurements made using a wide variety of experimental techniques (hence, accessing many orders of magnitude of nucleation rates) are included in the studies represented in these figures. Note also that the data in these figures encompass a wide range of saturation ratios.

Inspecting these figures, it is evident that the magnitude of the experimentally observed nucleation rate can differ dramatically from the prediction of standard CNT. For the four alcohols shown, CNT underpredicts the nucleation rate by up to six orders of magnitude at the low-temperature end of the plots and overpredicts the nucleation rate by up to six orders of magnitude at the high-temperature end. For n-nonane, CNT underpredicts the nucleation rate by up to nine orders of magnitude at the low-temperature end of the data and overpredicts it by up to four orders of magnitude at the high-temperature end. For water, a similar trend is observed, with the discrepancy equaling ± 4 orders of magnitude for measurements made with a wide variety of experimental techniques. For argon and nitrogen, the discrepancies between theory and experiment are even more severe. For argon, standard CNT underpredicts nucleation rates measured with a nucleation pulse chamber (10^5 to 10^9 cm^{-3} s^{-1}) by over

---

[8] Note that in calculating the nucleation rate predicted by CNT for water at temperatures below 0°C, it is assumed that the equilibrium vapor pressure equals the pressure of water vapor in metastable equilibrium with subcooled liquid water, as required by the capillarity model, although ice is the equilibrium condensed phase of water at these temperatures. A similar approach is taken for other substances at temperatures below their melting point.

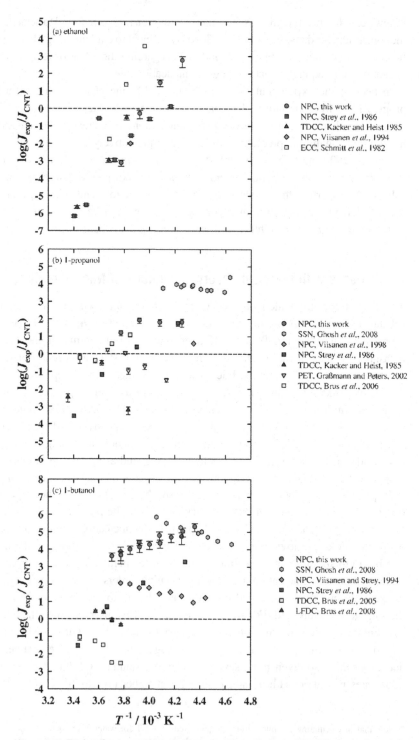

Figure 3.11 Ratio of measured nucleation rate to prediction of standard version of CNT, for 1-alcohols ($C_nH_{2n+1}OH$, $n=2$-4), from a number of studies. Acronyms in legend indicate experimental techniques: nucleation pulse chamber (NPC) (Manka et al. 2012; Strey et al. 1986;

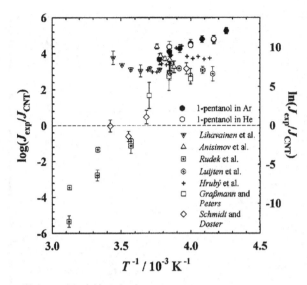

Figure 3.12 Ratio of measured nucleation rate to prediction of standard version of CNT, for 1-pentanol ($C_5H_{11}OH$), from a number of studies. Solid and open circles are from Iland et al. (2004), who explored the effect of carrier gas, either argon or helium. Other symbols in legend refer to Anisimov et al. (2000); Grassmann and Peters (2000); Hruby et al. (1996); Lihavainen et al. (2001); Luijten et al. (1997); Rudek et al. (1999); and Schmitt and Doster (2002). Reprinted from Iland et al. (2004) with the permission of AIP Publishing.

10 orders of magnitude and underpredicts nucleation rates measured using a supersonic nozzle (10^{16} to 10^{18} cm^{-3} s^{-1}) by about 20–30 orders of magnitude. For N_2, standard CNT underpredicts measurements made using a supersonic nozzle by about 15 orders of magnitude over the entire temperature range studied.

For each of the alcohols, n-nonane, and water, there is some narrow intermediate temperature range where standard CNT predicts the nucleation rate reasonably well. But, in general, this exercise clearly leads to the conclusion that CNT does a notably poor job of predicting the magnitude of the nucleation rate, for a wide variety of substances. Indeed, it does such a poor job that the fact that theory and experiment agree in some cases to within one or two orders of magnitude must probably be regarded as fortuitous.

3.11.5   Comparisons with Classical Theory: Effect of Temperature

If, as shown in Section 3.11.4, standard CNT does a poor job in general of predicting the magnitude of nucleation rates, Figures 3.11–3.16 appear to imply that it also fails to correctly predict the dependence of nucleation rate on temperature.

Figure 3.11 (cont.) Viisanen and Strey 1994; Viisanen et al. 1994, 1998), thermal diffusion cloud chamber (TDCC) (Brus et al. 2005, 2006; Kacker and Heist 1985), expansion cloud chamber (ECC) (Schmitt et al. 1982), supersonic nozzle (SSN) (Ghosh et al. 2008), piston expansion tube (PET) (Grassmann and Peters 2002), and laminar flow diffusion chamber (LFDC) (Brus et al. 2008a). Reprinted from Manka et al. (2012) with the permission of AIP Publishing.

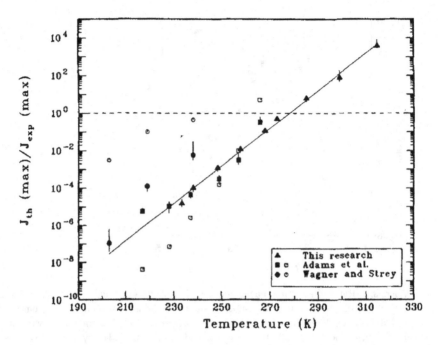

Figure 3.13 Ratio of theoretical nucleation rate, as predicted by standard CNT, to experimental measurements for *n*-nonane (C_9H_2O), from a number of studies. "This research" refers to Hung et al. (1989). Other symbols in legend refer to Adams et al. (1984), and Wagner and Strey (1984). Note that both ordinate and abscissa are inverted relative to the plots in Figures 3.11, 3.12, and 3.14–3.16. Reprinted from Hung et al. (1989) with the permission of AIP Publishing.

In some cases, such as in Figure 3.14 for water, the ratio of the experimental nucleation rate to the theoretical rate predicted by standard CNT appears to be reasonably well fit by an exponential function of inverse temperature, as indicated by the solid line in Figure 3.14. This line is given by

$$\frac{J_{\text{empirical}}}{J_{\text{CNT,standard}}} = \exp\left(-27.56 + \frac{6500}{T}\right), \tag{3.116}$$

where T is in Kelvin (Wölk and Strey 2001).

Thus the right-hand side of this equation represents an empirical correction factor by which the nucleation rate of water predicted by standard CNT can be multiplied to match the experimental data, to within the accuracy of the linear fit of Figure 3.14. This exercise is implemented in Figure 3.17, which shows the complete nucleation rate data from a compilation of measurements of 10 different studies of homogeneous nucleation of water, conducted over the period 1983 to 2010 (Manka et al. 2010). These studies employed a wide variety of experimental techniques, including laminar flow diffusion chamber (Manka et al. 2010), nucleation pulse chamber (Viisanen et al. 1993; Wölk and Strey 2001), laminar flow tube reactor (Mikheev et al. 2002), expansion wave tube (Holten et al. 2005; Luijten et al. 1997), expansion cloud chamber (Miller et al. 1983), thermal diffusion cloud chamber (Brus et al. 2008,

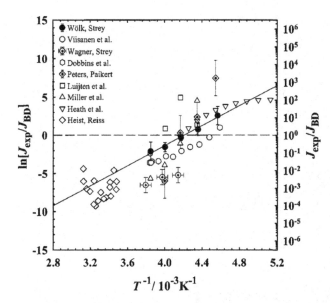

Figure 3.14 Comparison of water nucleation rates measured experimentally to the standard (Becker–Döring) version of CNT. Symbols in the legend refer to data of Dobbins et al. (1977), Heath (2001), Heath et al. (2002), Heist and Reiss (1973), Luijten (1998), Luijten et al. (1997), Miller (1976), Miller et al. (1983), Peters and Paikert (1989), Viisanen et al. (1993, 2000), Wagner and Strey (1981, 2001). Each data point represents the mean value of the ratio Jexp/JBD, taken over measurements at several values of saturation ratio along an isotherm. Error bars represent one standard deviation from the mean. Reprinted from Wölk et al. (2002) with the permission of AIP Publishing.

2009), and supersonic nozzle (Kim et al. 2004). Measurements span temperatures ranging from 210 to 320 K, saturation ratios from about 3 to 300, and nucleation rates ranging over 20 orders of magnitude, from about 10^{-2} to 10^{18} cm^{-3}·s^{-1}. The dashed lines in the figure show the prediction of the standard version of CNT multiplied by the right-hand side of equation (3.116).

Note that if instead the self-consistent version of CNT were used for this comparison, a different empirical correction factor would be deduced, in which the right-hand side of equation (3.116) is divided by e^{Θ}/S. As, by equation (3.28), $\Theta \equiv \sigma s_1/k_B T$, this factor tends to reduce the exponential temperature dependence in equation (3.116), but not by nearly enough, in general, to produce good agreement with experimental measurements. In the case of water, self-consistent CNT produces better agreement with experimental results at the high-temperature end of the experiments, but worse agreement at the low-temperature end.

Whether a relationship of the general form of equation (3.116) can be extended more generally to substances other than water is unclear. From Figure 3.11, it seems that ethanol does follow a similar trend, as does *n*-nonane in Figure 3.13. As for the other alcohols in Figures 3.11 and 3.12, the situation is unclear. There is perhaps weak agreement for these substances with the trend of equation (3.116), but some of the separate studies in these figures do not follow that trend or even contradict it, for

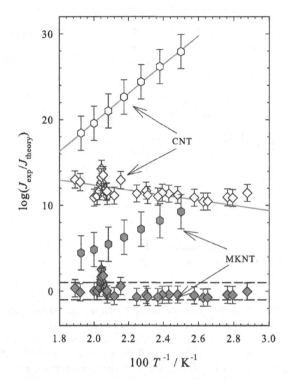

Figure 3.15 Measurements of argon nucleation rates compared to theoretical predictions of either the standard version of CNT (open symbols) or so-called "mean-field kinetic nucleation theory" (closed symbols) (Kalikmanov 2006). Diamond symbols refer to measurements using a nucleation pulse chamber (Iland et al. 2007), which accessed nucleation rates in the range 10^5 to 10^9 cm$^{-3}\cdot$s^{-1}. Hexagons refer to measurements using a supersonic nozzle (Sinha et al. 2010), which accessed nucleation rates of 10^{16} to 10^{18} cm$^{-3}\cdot$s^{-1}. Reprinted from Sinha et al. (2010) with the permission of AIP Publishing.

example, for 1-propanol, the nucleation pulse chamber data of Manka et al. (2012) and the piston expansion tube data of Grassmann and Peters (2002).

3.11.6   Comparisons with Classical Theory: Effect of Saturation Ratio

As seen in Sections 3.11.4 and 3.11.5, CNT does a poor job of predicting both the magnitude of the self-nucleation rate and the trend of its dependence on temperature. However, a considerable body of work in the literature asserts that CNT does at least do a good job of predicting the dependence of nucleation rate on saturation ratio.

For example, Hung et al. (1989), reviewing the three different studies of n-nonane nucleation on which Figure 3.13 is based, stated that "all three sets of data are consistent with the approximation that the supersaturation dependence is adequately fit by classical theory." Oxtoby, reviewing the measurements prior to 1992 of homogeneous nucleation rates of a number of different substances, stated that "The consensus of this work is that the variation of nucleation rate with supersaturation is well predicted by classical theory" (Oxtoby 1992). And equation (3.116), the empirical correction to standard

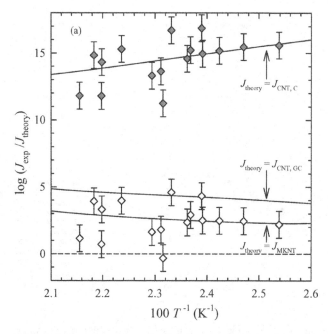

Figure 3.16 Experimental results for nitrogen nucleation in a supersonic nozzle, compared to theory. Notation: "CNC,C" = standard CNT; "GC" = self-consistent CNT (Girshick and Chiu 1990); "MKNT" = mean field kinetic nucleation theory (Kalikmanov 2006). Closed symbols = CNC,C. Open symbols = MKNT. For clarity no symbols are shown for CNT,GC. Reprinted from Bhabhe and Wyslouzil (2011) with the permission of AIP Publishing.

CNT proposed by Wölk and Strey (2001) to match the experimental results for water nucleation, is a function only of temperature, consistent with their claim in that work that the dependence of nucleation rate on saturation ratio is well predicted by CNT. In this section, we reexamine this assertion (Girshick 2014).

In Figure 3.10, for the 1-alcohols, and Figure 3.17, for water, the experimental isotherms are compared to the predictions of either standard CNT (Figure 3.10) or of standard CNT multiplied by the temperature-dependent correction, equation (3.116) (Figure 3.17). At first glance, it does appear that the measurements in both figures follow the trend predicted by CNT reasonably well regarding the dependence of nucleation rate on saturation ratio.

However, that conclusion, despite being often repeated in the literature, is flawed on two counts. First, almost always in nucleation experiments, experimental constraints limit the extent to which temperature and saturation ratio can be freely varied, independent of each other. Secondly, on closer inspection, the slopes of the experimental isotherms for the 1-alcohols in Figure 3.10, for water in Figure 3.17, and for other substances for which these types of plots have been reported do *not*, in general, agree well with the predicted saturation-ratio-dependence of CNT.

Regarding the first point, as discussed in Section 3.11.1, different experimental devices for studying nucleation examine characteristically different ranges of nucleation rate, and these tend to correspond to different ranges of temperature and saturation ratio.

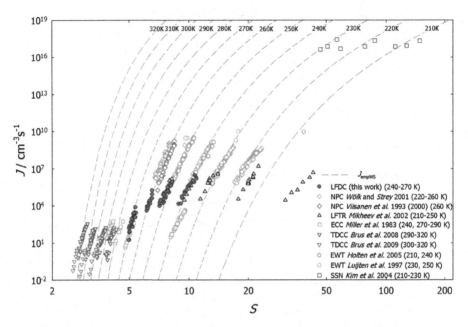

Figure 3.17 Compilation of measurements of homogeneous nucleation rates of water from 10 different studies, compared to predictions of standard version of CNT multiplied by the empirical temperature-dependent factor of Wölk and Strey (2001), given by equation (3.116) and shown by the dashed lines. Acronyms in legend refer to laminar flow diffusion chamber (LFDC) (Manka et al. 2010), nucleation pulse chamber (NPC) (Viisanen et al. 1993; Wölk and Strey 2001), laminar flow tube reactor (LFTR) (Mikheev et al. 2002), expansion cloud chamber (ECC) (Miller et al. 1983), thermal diffusion cloud chamber (TDCC) (Brus et al. 2008a, 2009), expansion wave tube (EWT) (Holten et al. 2005; Luijten et al. 1997), and supersonic nozzle (SSN) (Kim et al. 2004). Reprinted from Manka et al. (2010) with the permission of AIP Publishing.

This is illustrated for the case of water in Figure 3.18, which shows the ranges of nucleation rate, temperature, and saturation ratio accessed by eight different experimental methods (Wyslouzil and Wolk 2016). Clearly the ranges of temperature and saturation ratio explored by each apparatus are correlated: Measurements at low temperatures tend to involve relatively high saturation ratios, measurements at high temperatures tend to involve low saturation ratios, and measurements at intermediate temperatures tend to involve intermediate saturation ratios. This implies that Figure 3.14, which plots the ratio of the measured water nucleation rate to the rate predicted by standard CNT, versus inverse temperature, regardless of the saturation ratio, could instead have been plotted versus saturation ratio, regardless of the temperature. Had this been done, one could presumably construct an empirical correction factor for the CNT-predicted nucleation rate as a function of saturation ratio rather than of temperature. The range of the disagreement between experiment and standard CNT would be unchanged, that is, the vertical extent of the data in Figure 3.14 would remain the same, indicating a discrepancy of ±4 orders of magnitude, and, if the abscissa represented saturation ratio rather than inverse temperature, the plot would look rather similar. Thus there is no way to know from the experimental data whether the

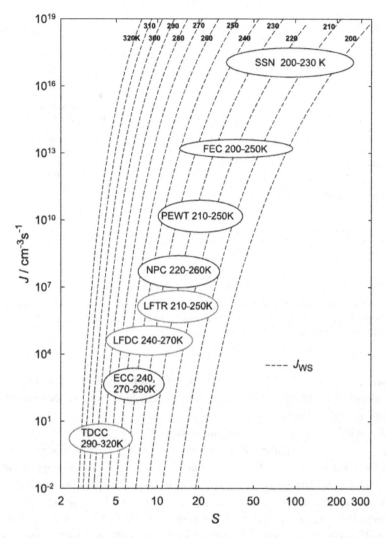

Figure 3.18 Experimental techniques that have been used to measure steady-state nucleation rates of water under various ranges of temperature, saturation ratio, and corresponding nucleation rate. From bottom to top, techniques include thermal diffusion cloud chamber (TDCC), expansion cloud chamber (ECC), laminar diffusion flow chamber (LFDC), laminar flow tube reactor (LFTR), nucleation pulse chamber (NPC), pulse expansion wave tube (PEWT), free expansion chamber (FEC), and supersonic nozzle (SSN). Dashed lines represent isothermal nucleation rates according to the empirical correlation by Wölk and Strey (2001), given by equation (3.116). Reprinted from Wyslouzil and Wölk (2016) (open access).

discrepancy between experiment and classical theory is due to an incorrect temperature dependence of the nucleation rate in the theory, an incorrect dependence on saturation ratio, or both.

Regarding the second point, which concerns the slopes of the experimental isotherms on a log–log plot of nucleation rate versus saturation ratio, the vast majority of such experiments have been done under conditions for which the slopes are steep (as in

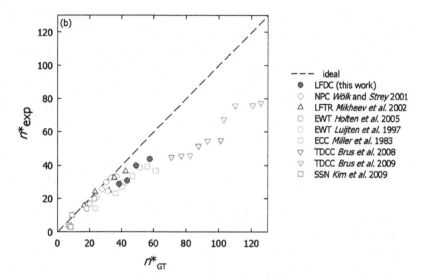

Figure 3.19 Values of critical cluster size n^* inferred from the nucleation theorem, applied to the experimental measurements of water nucleation rates shown in Figure 3.17, compared to the values predicted by the Gibbs–Thomson equation, equation (3.32). Reprinted from Manka et al. (2010) with the permission of AIP Publishing.

Figures 3.9, 3.10, and 3.17), and it may be difficult for the eye to discern differences in two slopes, both of which are steep, even though the values of these slopes may be quantitatively different.

However, a number of studies have carefully evaluated these slopes, including assessments of uncertainties, and have found that in general the slopes do not agree with the predictions of CNT. These studies typically evaluate the slope at the logarithmic midpoint of the range of saturation ratio for each isotherm. Actually, what the discussions of the results in these studies have focused on is not whether the dependence of nucleation rate on saturation ratio is well predicted by CNT (this question typically being dismissed by the assertion that it *is* well predicted) but rather on the experimental determination of the critical cluster size, n^*_{exp}. However, a *direct* measurement of the critical cluster size, based on its definition, equation (3.31), has rarely if ever been reported. Instead, the value of n^*_{exp} is inferred by means of the nucleation theorem. But as the nucleation theorem, equation (2.86), states that n^* is basically equivalent to $(\partial \ln J / \partial \ln S)_T$, the value of n^*_{exp} that is reported in these studies is simply a surrogate for the dependence of nucleation rate on saturation ratio.

Results from studies that conducted this exercise for various substances, including water, the 1-alcohols, argon, and nitrogen, are shown in Figures 3.19–3.22, respectively. Note that in each of these studies the nucleation theorem is written in the form given by equation (3.109). In other words, the value of n^*_{exp} shown in these figures is precisely the authors' determination from their experimental data of $(\partial \ln J / \partial \ln S)_T$, regardless of the slight deviation from that value given by equations (2.86), for self-consistent CNT, or (3.108), for the standard version of CNT. The abscissa on these

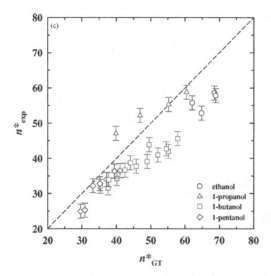

Figure 3.20 Values of critical cluster size n^* inferred from the nucleation theorem, applied to the experimental measurements of 1-alcohol nucleation rates shown in Figure 3.10, compared to the values predicted by the Gibbs–Thomson equation, equation (3.32). Reprinted from Manka et al. (2012) with the permission of AIP Publishing.

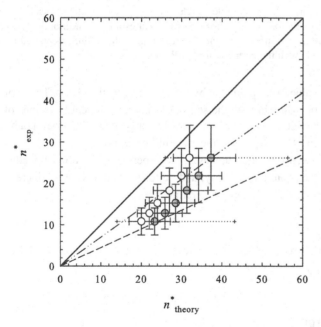

Figure 3.21 Values of critical cluster size n^* inferred from the nucleation theorem, applied to the experimental measurements of argon nucleation by Sinha et al. (2010), vs. theoretical values using either CNT (filled circles) or mean field kinetic nucleation theory (open circles) (Kalikmanov 2006). Solid line represents perfect agreement between experiment and theory. Dashed-dotted and dashed lines correspond to $n^*_{exp} = 0.7 \times n^*_{theory}$ and $n^*_{exp} = 0.45 \times n^*_{theory}$, respectively. Vertical error bars represent estimated uncertainty in nucleation rates. Horizontal lines represent estimated uncertainty in saturation ratios. Reprinted from Sinha et al. (2010) with the permission of AIP Publishing.

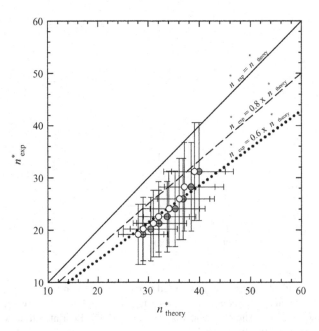

Figure 3.22 Values of critical cluster size n^* inferred from the nucleation theorem, applied to the experimental measurements of nitrogen nucleation in Bhabhe and Wyslouzil (2011), vs. theoretical values using either CNT (filled circles) or mean field kinetic nucleation theory (open circles) (Kalikmanov 2006). Vertical error bars represent estimated uncertainty in experimental nucleation rates; horizontal lines represent 5% error in the value of lnS. Reprinted from Bhabhe and Wyslouzil (2011) with the permission of AIP Publishing.

graphs gives the value of n^* predicted by CNT, that is, by the Gibbs–Thomson equation (3.32), which is the same in both the self-consistent and standard versions of CNT.

Inspecting these figures, one can conclude that in general CNT does a rather poor job of predicting $(\partial \ln J / \partial \ln S)_T$, with the possible exception of water for values of n^* below about 30. Even for water, however, the experimental and CNT-predicted values of $(\partial \ln J / \partial \ln S)_T$ deviate increasingly, and quite substantially, as n^* increases.

From equation (2.86), at given temperature,

$$J \propto S^{n^*+1}, \qquad (3.117)$$

so if n^*_{exp} is poorly predicted it follows that the exponent in the power-law dependence of J on S is poorly predicted.

3.12   Sources of Error

As discussed in Section 3.11.2, CNT does a reasonable job of predicting the critical supersaturation. Yet, while that is useful in its own right, it represents a quite insensitive test of the theory. As for the nucleation rate itself, comparisons between CNT and experimental results for a wide variety of substances indicate that CNT does a remarkably poor job of predicting the magnitude of the nucleation rate, as well as its dependence on both temperature and saturation ratio. Indeed, CNT seems so inaccurate that it is remarkable that it has remained such a robust theory for almost a century. It is still the primary touchstone for comparison with experimental studies, and it remains widely used in the numerical modeling of particle formation in many types of systems.

It thus remains of interest to address the possible sources of error in nucleation rates predicted by CNT. These possible sources of error can be divided into the following categories: uncertainties in the physical properties that are required as inputs to the CNT-predicted nucleation rate; errors in the mathematical approximations made in the development of CNT; errors in physical assumptions, other than the classical liquid droplet model; and lack of validity of the classical liquid droplet model.

3.12.1 Uncertainties in Physical Properties

As discussed in Section 3.7, the key physical properties of a substance that are needed to evaluate the steady-state nucleation rate predicted by CNT include the mass density $\rho(T)$ of the bulk liquid, the surface tension $\sigma(T)$ of a flat surface of the substance in equilibrium with its vapor, and the substance's equilibrium vapor pressure $p_s(T)$. To make the roles of these quantities explicit, let us rewrite equation (3.90) using the definitions of S and Θ, yielding

$$J = \frac{d_1^2}{6} \sqrt{\frac{2\sigma s_1}{m_1}} \frac{p_1 p_s}{(k_B T)^2} \exp\left\{ \frac{\sigma s_1}{k_B T} - \frac{4}{27} \frac{(\sigma s_1 / k_B T)}{[\ln(p_1/p_s)]^2} \right\}. \tag{3.118}$$

The mass density for most liquid substances is known with good accuracy. And, as long as one accepts the structureless liquid droplet model down to the size of a monomer, as done in CNT, then the quantities d_1 and s_1 follow directly from the mass density.

However, both the surface tension of the bulk liquid and the equilibrium vapor pressure may involve significant uncertainties, and errors in either or both of these quantities can translate into large errors in the nucleation rate predicted by equation (3.118).

Furthermore, each of the experimental methods for measuring nucleation rates has its own set of uncertainties associated with physical property inputs. For example, interpreting measurements from diffusion cloud chambers requires values for transport properties that are needed to calculate the profiles illustrated in Figure 3.8. It has been estimated that uncertainties in these properties, together with measurement uncertainties, could result in a total uncertainty in the nucleation rate at the reported values of temperature and saturation ratio of about one order of magnitude (Rudek et al. 1999). Brus et al. (2008b) reviewed the uncertainties associated with using any of several different expressions in the literature for these physical properties for water in the temperature range 290–320 K. They found that differences in the expressions used for equilibrium vapor pressure could translate to discrepancies of up to several tens of percent in nucleation rate.

3.12.2   Errors in Mathematical Approximations

The derivation of the steady-state nucleation rate in CNT makes several approximations of a purely mathematical nature. Although the distinction between mathematical and physical approximations is somewhat arbitrary, it manifests itself most clearly in the series of approximations made to evaluate the summation in equation (3.70), resulting finally in the expression for the nucleation rate given by equation (3.89). Equation (3.70) results from employing the physical model of CNT to evaluate the condensation rate constants k_n and the Gibbs free energies of cluster formation ΔG_n that appear in the general expression for the steady-state nucleation rate, equation (2.81). Proceeding beyond equation (3.70) then involves several mathematical approximations, as follows.

First, in equation (3.73), the summation over integer cluster sizes n ranging from 1 to M is replaced by an integral over a continuous size variable x that ranges from 0 to ∞.

Second, in the Taylor series expansion of $H(x)$ about a value of x for which $dH/dx = 0$ (equation (3.74)), terms higher than the second derivative term are neglected in equation (3.75).

Third, in equation (3.80), the lower limit of integration $-x'$ is replaced with $-\infty$ in equation (3.81).

Finally, in equation (3.86), the value of $\Theta(x')^{2/3}$ is assumed to be much greater than unity, allowing one to express x' by equation (3.87), which in turn leads directly to the expression for the nucleation rate given by equation (3.89).

One way to evaluate the total effect of these purely mathematical approximations is to compare the result of calculations that directly use equation (3.70) with those for the same conditions that instead use equation (3.89), where in evaluating the summation in equation (3.70) we count a sufficient number of terms for the summation to converge to within some desired tolerance.

From equation (3.70), one can write a dimensionless nucleation rate in CNT as

$$\frac{J}{\beta s_1 N_1} = \left\{ \sum_{n=1}^{M} \left[n^{2/3} S^{n-1} \exp\left[-\left(n^{2/3} - 1\right)\Theta \right] \right]^{-1} \right\}^{-1}. \tag{3.119}$$

The summand is in fact the function $e^{-H(n)}$ introduced in Section 3.8 in the process of converting the summation to an integral.

The summation on the right-hand side of this equation can be written term by term as

$$1 + \frac{\exp\left[\left(2^{2/3} - 1\right)\Theta\right]}{2^{2/3}S} + \frac{\exp\left[\left(3^{2/3} - 1\right)\Theta\right]}{3^{2/3}S^2} + \cdots + \frac{\exp\left[\left(M^{2/3} - 1\right)\Theta\right]}{M^{2/3}S^{M-1}}$$
$$= 1 + e^{-\Theta} \left[\frac{\exp\left[\left(2^{2/3}\Theta\right)\right]}{2^{2/3}S} + \frac{\exp\left[\left(3^{2/3}\Theta\right)\right]}{3^{2/3}S^2} + \cdots + \frac{\exp\left[\left(M^{2/3}\Theta\right)\right]}{M^{2/3}S^{M-1}} \right]. \tag{3.120}$$

Figure 3.23 shows the value of the right-hand-side of equation (3.119) as successive terms are added to the summation, using as examples a dimensionless surface tension of $\Theta = 10$ and saturation ratios of 5, 10, and 20, and compares these results to what one obtains from a nondimensionalized version of equation (3.89),

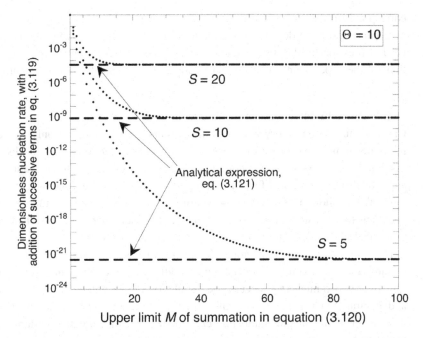

Figure 3.23 Value of the dimensionless nucleation rate given by equation (3.119), as successive terms are added to the summation, and comparison with the analytical expression, equation (3.121), for various saturation ratios and for a dimensionless surface tension $\Theta = 10$. The critical size n^* predicted by CNT for these three cases equals 11.0, 24.3, and 71.1, respectively, for saturation ratios of 20, 10, and 5.

$$\frac{J}{\beta s_1 N_1} = \frac{1}{3S}\sqrt{\frac{\Theta}{\pi}}\exp\left[\Theta - \frac{4}{27}\frac{\Theta^3}{(\ln S)^2}\right]. \tag{3.121}$$

As can be seen, the summation converges quite satisfactorily by the time the cluster size n is only slightly larger than the critical size, which, based on equation (3.32) for these three cases, equals 11.0, 24.3, and 71.1 for saturation ratios of 20, 10, and 5, respectively. Moreover, the values to which the summations converge are seen to be quite close to the values given by the analytical expression, equation (3.121), for nucleation rates that range in Figure 3.23 over 17 orders of magnitude.

Considering that several mathematical approximations were made in converting the summation to an integral to obtain the analytical expression, the agreement is impressive. We can thus conclude that these mathematical approximations cannot possibly explain the large discrepancies observed between experimental results and CNT.

3.12.3 Errors in Physical Assumptions, Aside from the Liquid Droplet Model

Classical nucleation theory makes a number of physical assumptions aside from the liquid droplet model. Among the more important of these assumptions are the following:

It is assumed that cluster growth and decay occur only by monomer addition and evaporation, neglecting cluster–cluster reactions; that the rate constant for condensation corresponds to the monomer flux to a flat, stationary surface; that monomer condensation and evaporation do not change the temperature of clusters; and, in comparing the CNT expression for the steady-state nucleation rate to experimental studies, that conditions of the experiments are such that steady-state nucleation is a valid assumption.

Neglect of Cluster–Cluster Reactions

Classical nucleation theory assumes that clusters grow and decay only by adding or removing monomers to/from the cluster, as given by reaction (2.9). As discussed in Section 2.1, this assumption is common to the theory of gas-phase nucleation, prior to any models being assumed for cluster properties and reactivities, and thus is not unique to CNT. Its validity rests on the observation that in many cases monomers are far more abundant than clusters of any size larger than the monomer, and thus clusters are far more likely to collide with monomers than with other clusters.

However, this scenario is not followed in all cases. In particular, there are two circumstances that could invalidate the monomer-only growth assumption: During the time over which nucleation occurs, the population of monomers may become depleted, and, contrary to CNT, the actual profile of $\Delta G_n(n)$ may exhibit local minima that lead to the existence of "magic numbers," cluster sizes that are relatively stable and that therefore become much more abundant than clusters of neighboring sizes.

Most models of nucleation neglect the fact that the population of monomers can be depleted by the clustering process itself, which can cause the assumption that monomers are far more abundant than clusters of any other size to become invalid. However, if one assumes that the system initially consists only of monomers (and possibly of an inert carrier gas), then at least the earliest stages of nucleation should be well described by monomer-only growth. Then, if monomers become significantly depleted by their attachment to growing clusters, the system could undergo a transition to a situation where cluster–cluster reactions, such as reaction (2.5),

$$A_2 + A_3 \rightarrow A_5, \tag{2.5}$$

become significant, in parallel with monomer attachment reactions.

One way to assess the importance of cluster–cluster reactions is by means of "brute force" numerical simulations that account for all such reactions, together with all single-monomer condensation and evaporation reactions, while still maintaining the classical liquid droplet model. Results of such numerical simulations, discussed in Section 3.13, indicate that under most conditions the cluster–cluster reactions have only a small effect on particle formation rates, justifying their neglect.

Secondly, while CNT models ΔG_n as a smooth, monotonically increasing function of n from the monomer up to the critical size (assuming $n^* \geq 2$), the true profile of $\Delta G_n(n)$ may deviate considerably from this behavior, resulting in cluster sizes that become relatively abundant. In such cases, the assumption that monomers are much more abundant than clusters of any other size may not be valid, and cluster–cluster reactions of the form of reaction (2.5) may be important. This phenomenon is discussed further in Chapters 6 and 7.

Errors in the Condensation Rate Constant

Classical nucleation theory models the condensation rate constant, discussed in Section 3.6.1, by assuming that clusters experience a flux of monomers that is given by the expression from gas kinetic theory for the flux of molecules in a gas at translational equilibrium to a plane. By further assuming that all such monomer–cluster collisions are effective in adding a monomer to the cluster, one obtains the CNT expression, equation (3.54), for the forward rate constant k_n.

If one compares the CNT expression for the rate constant to the "exact" value of k_n derived from hard-sphere gas kinetic theory for collisions between monomers and clusters, one finds a substantial discrepancy, given by equation (3.55) and graphed in Figure 3.5. For cluster sizes smaller than $n = 10$, the CNT approximation is seen to underpredict the collision rate by more than a factor of two, and even for $n = 100$ the approximation is low by about one-third. Considering the role of k_n in the summation expression for the steady-state nucleation rate, for example, equation (2.80), it follows that this discrepancy causes a significant underprediction in the nucleation rate. This error is not enough to account for orders-of-magnitude discrepancies between CNT and experiment, but neither is it negligible.

Furthermore, CNT effectively assumes that the monomer–cluster collision rate equals the condensation rate and that the collision rate is unaffected by intermolecular forces between the monomer and cluster prior to the moment of collision. Depending on the substance and on conditions, a variety of effects might render these assumptions dubious. For example, in Section 6.5.2 we consider the need for a third-body collision partner to stabilize a cluster that has added a monomer, an effect that may be important at the smallest cluster sizes and at low total gas pressures. In Section 6.5.3, we consider the effect of attractive intermolecular forces on condensation rate constants. The combined effect of CNT's simplistic model for the monomer flux to a cluster together with its neglect of the effect of intermolecular forces on collision rates is shown in Figure 6.9, for the example of aluminum clusters, based on atomistic simulations by Li and Truhlar (2008).

Additionally, one must acknowledge the possibility that, depending on the substance, n-mers and $(n + 1)$-mers may be chemically distinct structures and that the stepwise reaction that adds a monomer to an n-mer may actually involve a chemical reaction not merely a physisorption. In that case, the stepwise reaction might involve an energy barrier such that only sufficiently energetic collisions result in monomer addition to the cluster. That, together with geometric factors affecting the probability that a collision results in a reaction, could cause the actual value of k_n to be orders of magnitude lower than in the CNT model, which would lead to errors in predicted nucleation rates in the direction opposite to the effects discussed above in this section.

It is also worth noting that the comparisons of experimental nucleation rates to CNT predictions, shown in Figures 3.11–3.16, while indicating that the discrepancies between CNT and experiment are indeed quite large, do not in general indicate a preferred direction for the discrepancy, that is, whether CNT underpredicts or overpredicts the nucleation rate.

Isothermal Condensation and Evaporation

In the CNT model, a cluster is a liquid droplet that grows by the addition of monomers from the vapor phase. Thus monomer addition is explicitly a condensation event. However, CNT does not consider the implication of this model that condensation heats the cluster by releasing the enthalpy of vaporization. Likewise, in CNT the reverse reaction is explicitly treated as evaporation, yet CNT does not account for evaporative cooling of the cluster. Instead, CNT assumes that all clusters and gases in the system are in thermal equilibrium, that is, at the same temperature. Yet many of the key inputs to the CNT model, for example, surface tension, depend on temperature.

If a cluster experiences temperature change due to condensation or evaporation, it would tend to be brought back into thermal equilibrium with the gas that surrounds it by collisions with the gas molecules. The time required for such thermalization with the gas depends on the gas total pressure and chemical composition. Therefore, if this effect is important, one can expect nucleation rates to depend on carrier gas pressure and composition, although these factors do not appear in CNT.

A number of studies have examined this isothermal assumption and have developed various approaches to considering nonisothermal nucleation (Campagna et al. 2020; Feder et al. 1966; Ford and Clement 1989; Grinin and Kuni 1989; Kantrowitz 1951; McGraw and Laviolette 1995; Salpeter 1973; ter Horst et al. 2011; Wyslouzil and Seinfeld 1992; Yang et al. 2019). Studies that have explored the effect on nucleation rates of total gas pressure are reviewed by Brus et al. (2006), while studies on the effect of carrier gas composition are reviewed by Wyslouzil and Wölk (2016). Many though not all of the experimental studies have confirmed that carrier gas pressure and composition affect nucleation rates, with the strongest effects generally found at the lowest pressures and with carrier gases such as helium. These are indeed the conditions under which one would expect cluster thermalization by collisions with background gas molecules to be least effective.

In most of the cited work, the experimental observations have usually been explained in terms of classical theory, relaxing the assumption of isothermal conditions while maintaining the presumption that cluster growth and decay occur by monomer condensation and evaporation, often in conjunction with the assumption that the critical-size cluster is large enough to be treated as a structureless liquid droplet, and that only clusters of critical size matter in the nucleation rate.

It should also be noted, however, that the observation of carrier gas pressure and composition effects does not prove the correctness of classical theory. From an atomistic viewpoint, many reactions of the type presumed to constitute the cluster growth sequence, reaction (2.9), require a third body to stabilize the excited complex formed as a transition state to the final product of the reaction, as discussed in Section 6.5.2. Moreover, most chemical reactions are either endothermic or exothermic, implying that the population of molecules formed by such reactions may or may not be thermally equilibrated with a background gas, typically with some characteristic time required for such thermalization. This characteristic time indeed depends on the degree of nonequilibrium and on the pressure and composition of the background gas.

Validity of the Steady-State Assumption

Experimental studies of nucleation almost universally assume steady-state nucleation. This assumption is justified by the belief that the time lag required to achieve steady state is much shorter than characteristic times for changes in the conditions that drive nucleation.

From analyses based on CNT for a range of conditions, the time lag required to achieve steady-state nucleation is found to lie in the range of a fraction of a microsecond to about 100 microseconds. That is indeed quite short compared to the timescales for changes in conditions that drive nucleation in many types of physical systems. This observation undoubtedly underlies the often unquestioning assumption that steady-state nucleation exists.

However, there are at least two reasons for questioning the validity of the steady-state assumption in some cases. First, real systems do exist where the timescales for change in the conditions that drive nucleation are themselves on the order of microseconds. One example of such a system is supersonic nozzle expansions, which are an important tool for studying nucleation.

Secondly, and more fundamentally, as estimates of the time lag required to achieve steady-state nucleation have been based on CNT, these estimates are subject to the same possibly large inaccuracies as CNT itself. This situation is examined in Chapter 7, where it is seen that transient analyses based on atomistic calculations of free energies of cluster formation indicate time lags required to achieve steady-state nucleation that may be several orders of magnitude longer than in CNT.

3.12.4   Validity of the Liquid Droplet Model

The most fundamental source of discrepancy between nucleation rates predicted by CNT and measured nucleation rates is likely the liquid droplet model itself. In particular, and as discussed in more detail in Chapter 6, the free energy of cluster formation estimated under the capillarity approximation may differ substantially from its actual value, both in magnitude and in its functional form, $\Delta G_n(n)$. The nucleation rate is particularly sensitive to the value of ΔG_n, as it appears in the argument of the exponential in equations (2.80) and (2.81). Classical nucleation theory treats clusters as structureless spheres that can be characterized as having a surface tension corresponding to that of the bulk liquid of the same substance, in equilibrium with its vapor. Consequently, the stepwise free energy change that accompanies cluster growth by the addition of one monomer is modeled in terms of the work against surface tension that is required to stretch a liquid surface, resulting in equation (3.23),

$$\Delta G_{n-1,n}(p_s) = \sigma(s_n - s_{n-1}). \tag{3.23}$$

The application of this relation down to the smallest cluster sizes, even down to dimerization by collision and sticking of two monomers, is obviously a crude approximation. Indeed, a large body of work indicates that small clusters of most substances are far from being structureless spheres and that the free energy change required to evaporate a monomer from a cluster may be highly size dependent and thus not well described in terms of a macroscopic liquid surface tension.

It is the stepwise free energy change given by equation (3.23), rather than the expression for the total Gibbs free energy of cluster formation ΔG_n, that should be regarded as the essence of the classical model. In the limit, as a cluster grows large enough to approach bulk behavior, it can be expected to exhibit macroscopic surface tension, and thus the stepwise free energy change would be expected to approach the value given by equation (3.23). However, it does not follow that the *total* free energy change ΔG_n predicted by the classical model should be expected to approach that of a liquid droplet of the same size. This is because ΔG_n is the sum of all the stepwise changes involved in growing from a monomer to an n-mer, as given by equation (3.24). Thus errors in the stepwise free energy change up to size n are cumulative in ΔG_n, notwithstanding that some of the errors may be of opposite sign and thus might tend to offset each other. Therefore, if the stepwise free energy change given by the liquid droplet model is a crude approximation, then ΔG_n itself can be expected to be an even worse approximation. Considering the key role played by ΔG_n in the summation expression for the nucleation rate, equations (2.68), (2.80), and (2.81), it is then not surprising that the magnitude of the nucleation rate is poorly predicted by CNT.

For the same reason, the discrepancy between CNT and experimental measurements regarding the temperature dependence of the nucleation rate is also to be expected. In general, Gibbs free energy depends on temperature, and the specific heats of clusters of different sizes may differ, meaning that their enthalpies, entropies, and Gibbs free energies may exhibit temperature dependencies that differ from one another as well as from that of the macroscopic surface tension.

However, CNT's poor prediction of the dependence of nucleation rate on saturation ratio, discussed in Section 3.11.6, is at first glance surprising. Comparing equations (2.80) and (2.81), it is clear that the appearance of the S^{n-1} term in equation (2.81) is purely a consequence of the dependence of the Gibbs free energy of an ideal gas on pressure, as given by equation (2.76). This is straightforward thermodynamics and can hardly be the source of the discrepancy between CNT and experimental results. Likewise, the $N_1{}^2$ prefactor in equation (2.81), which could instead be written as $[SN_s(T)]^2$, is a straightforward result of kinetics, as discussed in Section 2.7, and really does not denote a dependence on saturation ratio per se.

Rearranging equation (2.81) and expanding the summation term by term,

$$J = \left\{ \frac{1}{k_2 N_1{}^2} + \frac{\exp[\Delta G_2(p_s)/k_B T]}{k_3 N_1{}^2 S} + \frac{\exp[\Delta G_3(p_s)/k_B T]}{k_4 N_1{}^2 S^2} + \cdots + \frac{\exp[\Delta G_M(p_s)/k_B T]}{k_{M+1} N_1{}^2 S^{M-1}} \right\}^{-1}.$$

(3.122)

The argument of the exponential in each term depends only on temperature; the $N_1{}^2$ term arises from straightforward kinetics (two-body collision rates scale on the product of the collision partners' number densities); and the S^{n-1} term arises from straightforward thermodynamics (the dependence of Gibbs free energy on pressure). Thus if, as seems to be the case, every term in the series has a correct dependence on saturation ratio, independent of the model assumed for $\Delta G_n(p_s)$, it is at first puzzling to observe that experimental results do not predict a correct saturation-ratio dependence of CNT. Note that in the first term on the right-hand side of equation (3.122) we have utilized $\Delta G_1 = 0$, as done correctly in the self-consistent version of CNT, whereas the standard version violates this requirement. However, as noted in Section 3.11.6, the resulting error in the standard version is not nearly large enough to explain the discrepancies between theory and experiment for the critical size, and hence the dependence on saturation ratio.

The resolution of this conundrum lies in noting that, even though each individual term in the summand in equation (2.81) has a correct dependence on saturation ratio, which is unaffected by the model assumed for $\Delta G_n(p_s)$, the saturation-ratio dependence *of the sum itself* is affected by the model assumed for $\Delta G_n(p_s)$. That is because the sum is typically dominated by terms close to the critical size, where $\Delta G_n(p_1)$ has its maximum value, and the location of this maximum does indeed depend on the model chosen for $\Delta G_n(p_s)$.

One way to illustrate this is to again consider the limiting case where the summation is strongly dominated by the single largest term, for which $n = n^*$, as in equation (2.84), which can be written as

$$J_{1-\text{term}} = k_{n^*+1} N_s(T) S^{n^*+1} \exp\left[-\frac{\Delta G^*(p_s)}{k_B T}\right]. \tag{3.123}$$

This expression emphasizes that the saturation-ratio-dependence of nucleation rate is directly tied to the value of the critical size and that all of the discrepancy between CNT and the saturation-ratio-dependence of measured nucleation rates arises from errors in $\Delta G_n(p_s)$ that cause the critical size to be incorrectly predicted, even though $\Delta G_n(p_s)$ itself depends only on temperature.

Aside from the fact that the classical liquid droplet model can be expected to make quantitative errors in ΔG_n, the functional form assumed for $\Delta G_n(n)$ may be qualitatively incorrect. In the classical model, $\Delta G_n(p_s)$ is a monotonically increasing function of n, because the stepwise free energy change given by equation (3.23) is positive for all values of n. Consequently, within the condensation/evaporation regime, $\Delta G_n(p_1)$ is a smooth, unimodal function of n, with a single maximum that defines the critical size, as illustrated in Figure 3.3.

In reality, however, for many substances clusters have a size-dependent structure, and clusters of some sizes (so-called "magic numbers") may be more thermodynamically stable at pressure p_1 than those of other sizes, implying that they occupy a local minimum in Gibbs free energy space, $\Delta G_n(p_1)$. Thus for many substances the true situation for the free energy of cluster formation $\Delta G_n(p_1)$ may be as illustrated hypothetically in Figure 3.24, which is fundamentally different than the smooth unimodal curve that results from the liquid droplet model. If this seems far-fetched, that is only because the liquid droplet model is so entrenched in our way of thinking about nucleation, in spite of the fact that it lacks proof and fails to provide correct predictions of nucleation rates. Indeed, as shown in Chapter 6, high-level computational chemistry calculations for specific substances indicate that the type of free energy landscape illustrated in Figure 3.24 may be qualitatively common, even for such a familiar substance as water.

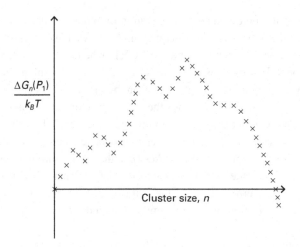

Figure 3.24 Hypothetical graph of atomistic values of Gibbs free energy of cluster formation versus cluster size, for a value of the saturation ratio lying in the condensation/evaporation regime.

If this is so, then consider the implications for the nucleation rate as expressed in the various forms of the summation expression, equations (2.68), (2.80), and (2.81). If $\Delta G_n(p_1)$ has a functional form as in Figure 3.24, rather than as in Figure 3.3, then multiple kinetic bottlenecks to cluster growth exist. At steady state, an overall critical size could still be identified as the location of the global maximum, but its value might be quite different than in the liquid droplet model, implying a quite different dependence of the nucleation rate on saturation ratio, as given by the nucleation theorem. This point is discussed in more detail in Chapter 6. The implications of having multiple kinetic bottlenecks for the time required to achieve steady-state nucleation are discussed in Chapter 7.

3.13   Comparison of CNT with Discrete Numerical Simulations

We conclude this chapter with a comparison of CNT to what might be termed "brute force" numerical simulations that solve population balance equations for clusters of each discrete size by calculating condensation and evaporation rates for each stepwise reaction, up to some size that is larger than the critical size. Larger cluster sizes are then treated with a more approximate model, either by integrating over a continuous particle size distribution function (Gelbard and Seinfeld 1979); or with a moment-type model, where one solves for the first few moments of a particle size distribution function (Friedlander 1983); or with a "sectional" model, in which larger particles are treated by binning them into sections of finite width in particle-size space (Gelbard et al. 1980; Girshick et al. 1990; Rao and McMurry 1989; Wu and Flagan 1988).

Such models can be characterized as being fully kinetic, as they calculate forward and reverse rates for all reactions of the form of reactions (2.9) and (2.5), while making no a priori assumptions regarding the nucleation rate. However, the cited models still utilize a central feature of classical theory – namely, the forward and reverse rate constants for monomer condensation/evaporation are related by the CNT expression for the stepwise free energy change, $\Delta G_{n-1,n}$, as given by equations (3.22) and (3.23).

For example, Girshick et al. (1990) reported numerical simulations using a discrete-sectional model and compared the results that assumed the steady-state nucleation rate given either by the self-consistent version of CNT, equation (3.90), or the standard version, from equation (3.107). They assumed an initial value for the saturation ratio and allowed monomers to be depleted by condensation.

For monomers, the population balance equation can be written as

$$\frac{dN_1}{dt} = -N_1 \sum_{n=1}^{\infty} \beta_{1,n} N_n + \sum_{n=2}^{\infty} (1 + \delta_{2,j}) E_n N_n, \tag{3.124}$$

while for clusters of all sizes larger than the monomer one has

$$n \geq 2: \quad \frac{dN_n}{dt} = \frac{1}{2} \sum_{i+j=n} \beta_{i,j} N_i N_j - N_n \sum_{i=1}^{\infty} \beta_{i,n} N_i + E_{n+1} N_{n+1} - E_n N_n. \tag{3.125}$$

In these equations, $\beta_{i,j}$ is the collision frequency function, given by equation (3.45), and E_n is the evaporation coefficient, given, consistent with equation (3.59), by

$$E_n = \beta_{1,n-1} N_s \exp\left\{ \Theta \left[n^{2/3} - (n-1)^{2/3} \right] \right\}. \tag{3.126}$$

The first term on the right-hand side of equation (3.124) represents the loss of monomers by condensation to clusters of all sizes. The second term represents the generation of monomers by evaporation from clusters of all sizes larger than the monomer. Here $\delta_{2,n}$, the Kronecker delta, accounts for the fact that evaporation from a dimer creates two monomers. It is defined by

$$\delta_{2,n} = \begin{cases} 1, \text{if } n = 2, \\ 0, \text{otherwise.} \end{cases} \tag{3.127}$$

The terms on the right-hand side of equation (3.125) represent, respectively, creation of n-mers by collision of two clusters whose sizes sum to n; loss of n-mers by collisions with clusters of all other sizes; and gain and loss of n-mers by evaporation of one monomer. The first two of these terms are called "coagulation," with the exception that, if one of the collision partners is a monomer, it is generally termed condensation. The factor ½ in front of the first summation avoids double-counting collisions, similar to the factor $1/\zeta$ in the expression for the bimolecular collision rate, equation (3.37).[9]

Note that the forward and reverse rate constants in equations (3.124) and (3.125) utilize the "exact" collision frequency function from ideal gas kinetic theory, rather than the more approximate version of CNT, which instead utilizes the flux of monomers to a plane, as discussed in Section 3.6.1. However, the forward and reverse rate constants are still related to each other by equation (3.57), using the identical stepwise free energy

---

[9] Note that in the first term on the right-hand side of equation (3.124), $\beta_{1,1} N_1 N_1$, which represents the monomer–monomer collision rate, is not multiplied by ½. The reason is that each such sticking collision removes two monomers.

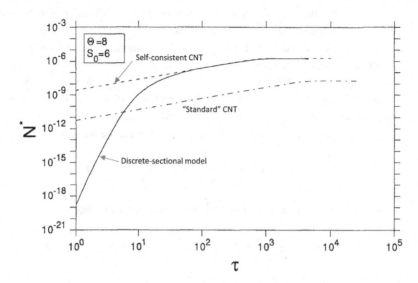

Figure 3.25 Temporal evolution of the number of particles larger than the critical size, calculated by three methods described in Section 3.13. Dimensionless number density N^* and dimensionless time τ are defined by equations (3.128) and (3.129), respectively. Adapted from Girshick et al. (1990) with the permission of Taylor & Francis Ltd., www.tandfonline.com.

change as in CNT, equation (3.22). Thus the essence of the liquid droplet model is preserved, while several important assumptions are relaxed.

Girshick et al. (1990) solved these equations for clusters of each discrete size up to $n = 180$, while similar population balance equations were solved for particles of larger sizes by grouping them into 20 sections, logarithmically spaced by particle volume. Figure 3.25 shows results of such a simulation, for a case where the initial saturation ratio equals 6, and the dimensionless surface tension equals 8. For these conditions, the critical size given by equation (3.32), based on the initial saturation ratio, equals 26.4. The graph plots the dimensionless number density, N^*, of all "stable" particles – particles larger than the critical size – versus dimensionless time, τ, where these are defined by

$$N^* \equiv \frac{N}{N_1(0)}, \tag{3.128}$$

N here being the total number density of stable particles, while $N_1(0)$ denotes the initial value of the number density of monomers and

$$\tau \equiv N_1(0)\beta_{1,1}t. \tag{3.129}$$

In Figure (3.25), the results of the discrete-sectional calculation are compared to two other models, which assumed the steady-state nucleation rate given either by self-consistent CNT or by standard CNT, with each being coupled to the moment-type model of Friedlander (1983), which accounts for coagulation.

As seen in Figure 3.25, at early times the results of the three models differ considerably. This is because the initial number density of stable particles in the discrete-sectional model equals zero, as finite time is required for clusters to grow from monomers up to the critical size, while the other two models immediately start generating stable particles at the steady-state nucleation rates they assume. (For a discussion of this "transient nucleation" period, before steady-state nucleation is achieved, see Chapter 7.) However, by $\tau \approx 100$, the results of the discrete-sectional model agree excellently with the results of the model that assumes the steady-state nucleation rate given by self-consistent CNT. Meanwhile, consistent with equation (3.107), the model that assumes the nucleation rate given by standard CNT is lower by the factor e^{Θ}/S, which in this case, based on the initial value of S, equals 497. Here it is interesting to note that the self-consistent and standard versions of CNT both assume the same stepwise free energy change for monomer condensation/evaporation, equation (3.22), as does the discrete-sectional model. The discrepancy between the discrete-sectional calculation and the standard version of CNT lies entirely in the self-consistency error that standard CNT makes in assigning a value of ΔG_n, as discussed in Section 3.10.

In all three models in Figure 3.25, it is only nucleation that generates stable particles, whereas coagulation of stable particles with each other reduces the number of stable particles. However the flattening of the three curves at later times can be ascribed mainly to monomer depletion, which relieves the supersaturation, reducing nucleation rates. Coagulation was found in this work to be important only at much higher saturation ratios (not shown in the figure), because these lead to the formation of much higher particle number densities, and coagulation rates scale on the square of number densities. Thus, the excellent agreement, once steady-state nucleation is achieved, between the discrete-sectional model and self-consistent CNT can be ascribed entirely to their agreement on the steady-state nucleation rate.

These results are instructive. The discrete-sectional model uses the "exact" kinetic theory collision rate between monomers and clusters, accounts for cluster–cluster reactions and monomer depletion, does not assume the existence of a steady-state nucleation rate, and makes none of the mathematical approximations that enter the derivation of the CNT expression for the steady-state nucleation rate. In contrast, CNT models monomer–cluster collision rates using the monomer flux to a plane; ignores cluster–cluster reactions; assumes the existence of steady-state nucleation; and makes a number of mathematical approximations in converting the summation expression for the steady-state nucleation rate to a simple analytical expression.

Therefore, the excellent agreement seen in Figure 3.25 between the discrete-sectional model and the (self-consistent) CNT nucleation rate is impressive validation of the work done by the pioneers who developed CNT, long before computational tools were available that could conduct such "brute force" numerical simulations as in a discrete-sectional model.

On the other hand, one can again conclude that the notably poor agreement between nucleation rates predicted by CNT and those measured in the large body of experimental studies that have been conducted, for a large variety of substances and conditions, is mainly ascribable to the fundamental error in the liquid droplet model, as embodied in the stepwise free energy change. It can be anticipated that this error can be rectified only by adopting atomistic approaches to predicting cluster free energies, as discussed in Chapter 6.

Homework Problems

3.1 Examine the inequality shown in equation (3.21) for the case of water at 300 K and a saturation ratio of 10. Can you identify conditions under which the inequality would not be valid, and where one would still be in the condensation/evaporation regime?

3.2 Equation (3.30) gives an expression for the saturation ratio S^* that constitutes the lower boundary for the "collision-controlled" regime, treating cluster size n as a discrete quantity. Find an expression for S^* if cluster size is treated instead as a continuous variable x.

3.3 Consider a gas consisting of 5% (by mole fraction) water vapor in helium at a total pressure of 1 bar and temperature of 340 K that is expanded through a nozzle isentropically to a downstream pressure of 0.6 bar. You can assume that the He–H_2O mixture remains well mixed and that the specific heat of the mixture is given by the specific heat of helium.

Referring as needed to a table of properties of saturated water vapor, find the saturation ratio S of water vapor at the downstream conditions.

3.4 Figure 3.3 plots $\Delta G_n / k_B T$ for a dimensionless surface tension of $\Theta = 10$ and for several values of saturation ratio. Prepare a similar figure for water at a temperature of 290 K, and for saturation ratios of 10, 20, and 100. Obtain needed property data from an appropriate reference.

What is the value of Θ for water at 290 K? What are the critical sizes (integer form n^*) at each of the saturation ratios?

Show your work that you use to prepare your graph.

3.5 Using the Kelvin equation, prepare a plot of the vapor pressure of mercury that would exist at equilibrium at 42°C over mercury droplets with diameters ranging from 1 nm to 100 nm.

Properties of mercury at 42°C: $\rho = 13.5 \text{ g/cm}^3$; $\sigma = 484 \text{ mN/m}$; $p_s = 1.0 \text{ Pa}$.

3.6 Based on self-consistent CNT, calculate the steady-state nucleation rate of water at these conditions of temperature and saturation ratio:

(a) $T = 300 \text{ K}, S = 3.5$

(b) $T = 250 \text{ K}, S = 9$

(c) $T = 210 \text{ K}, S = 100$.

3.7 For the three cases in the previous problem, calculate the critical size (integer value n^*) based on CNT. Then, using the nucleation theorem, use linear extrapolation (in terms of $\ln J$ vs. $\ln S$) to extend your predicted nucleation rates to saturation ratios slightly above and below the values assumed in each case: In (a), extrapolate your results to the range $S = 3 - 4$; in (b) to the range $S = 8 - 10$; and in (c) to the range $S = 90 - 110$. Show your predicted nucleation rates on a graph that is of the same form as Figure 3.17. Compare your results (approximately) to the experimental data shown in Figure 3.17, and comment on discrepancies with regard both to the magnitude of nucleation rate and to the dependence of nucleation rate on saturation ratio.

3.8   In Figure 3.10, it can be observed that as temperature decreases the saturation ratio required to achieve a given water nucleation rate increases. Can you give a physical explanation for this?

3.9   For the same conditions as assumed in Figure 3.23, prepare a similar graph, but showing the value of each separate term in the summand in equation (3.119), with $M = 100$, rather than the value of the summation itself, for saturation ratios of 5, 10, and 20. What does this graph tell us about the validity in CNT of the one-term approximation for the nucleation rate, equation (2.84)?

4 Classical Theory of Multicomponent Nucleation

In the previous chapter, we considered classical nucleation theory, applied to the scenario in which nucleation occurs via condensation of a supersaturated vapor of a single chemical substance. Classical nucleation theory (CNT) has also been applied to situations where vapors of two or more different substances condense to form multi-component clusters. Depending on the number of substances that condense, this is known as binary, ternary, etc., nucleation, and in general as multicomponent nucleation.

4.1 General Considerations

In applying CNT to multicomponent nucleation, several key assumptions are the same as in CNT for single-component nucleation. Clusters are modeled as spherical liquid droplets, having the same mass density as the bulk liquid of the same chemical composition and same temperature, and having the same surface tension as a flat surface of the liquid, in equilibrium with its vapor, again at the same temperature. Cluster growth is assumed to occur by single monomer addition only, and cluster decay occurs by monomer evaporation only.

Multicomponent nucleation can occur even when none of the condensing vapors is supersaturated with respect to its own bulk condensed phase. Rather, two or more components in the vapor mixture may be supersaturated with respect to a multicomponent condensed phase containing the same substances.

The most studied example is the homogeneous nucleation of particles in the earth's atmosphere, which is believed in most cases to involve either binary nucleation of sulfuric acid together with water or various organic molecules; or ternary nucleation involving an acid (especially sulfuric acid), a base (especially ammonia), and water.

The classical theory of multicomponent nucleation builds on the liquid droplet model of CNT to treat multicomponent clusters as droplets of ideal liquid solutions. For binary nucleation, the theory was first proposed by Flood (1934) and further developed by Reiss (1950).

The elegant simplicity of single-component homogeneous nucleation allows CNT to derive a simple analytical expression, equation (3.90), or equivalent forms, for the steady-state nucleation rate as a function of temperature, saturation ratio, and bulk substance properties. Unfortunately, the more complicated scenario of multicomponent nucleation does not admit of such a simple closed-form expression for the nucleation rate. Rather, the approach taken in most of the literature is to emphasize the role of the critical cluster and to utilize equations (3.91) or (3.94) to predict the steady-state nucleation rate.

Here we present a brief introduction to the topic. For more detailed treatments, the reader is referred to the textbook by Vehkamäki (2006) and to the recent review by Elm et al. (2020). In the following, we focus for simplicity on binary nucleation. The extension to nucleation involving three or more substances is straightforward.

4.2   Gibbs Free Energy of Formation of a Binary Cluster

Consider a gas mixture that includes two condensible vapors, A and B, that combine to form clusters containing n molecules of A and m molecules of B. The overall reaction to form a cluster can be written as

$$n\,A + m\,B \rightarrow A_n B_m, \tag{4.1}$$

as illustrated in Figure 4.1.

Many sequences of single-monomer addition events are possible that would result in the overall reaction given by reaction (4.1). Regardless of the sequence, the change in Gibbs free energy associated with this overall reaction can be written as

$$\Delta G_{nm} = G_{nm} - n\tilde{G}_{A,v} - m\tilde{G}_{B,v}, \tag{4.2}$$

where the subscript "nm" denotes the cluster $A_n B_m$, and $\tilde{G}_{A,v}$ and $\tilde{G}_{B,v}$ are the free energies per molecule of vapor phase A and B, respectively, all at a common temperature T.

In single-component nucleation, ΔG_n depends on the number n of monomers comprising the cluster, on the partial pressure p_1 of the monomer vapor, and on temperature. In binary nucleation, ΔG_{nm} depends on the number of monomers of both species that comprise the cluster, as well as on the partial pressures in the gas phase of both A and B, denoted p_A and p_B, respectively, and on temperature. Thus,

$$\Delta G_{nm} = \Delta G_{nm}(n, m, p_A, p_B, T). \tag{4.3}$$

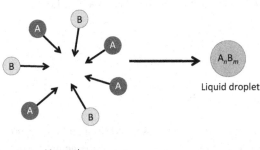

Vapor phase
n monomers of A
m monomers of B

Figure 4.1 Overall reaction to form an $A_n B_m$ cluster from vapor-phase monomers of A and B. The cluster is modeled in classical theory as a spherical liquid droplet.

The cluster $A_n B_m$ is regarded in the classical model as a spherical liquid droplet consisting of a two-component ideal solution. As in the classical model for single-component nucleation, the free energy of the droplet can be written as the sum of volume and surface terms:

$$G_{nm} = n\tilde{G}_{A,l}(p_A) + m\tilde{G}_{B,l}(p_B) + \sigma s_{nm}, \qquad (4.4)$$

where $\tilde{G}_{A,l}$ and $\tilde{G}_{B,l}$ are the free energies per molecule of A and B, respectively, in their pure liquid phases, s_{nm} is the surface area of a droplet containing n A molecules and m B molecules, and σ is the surface tension of a flat surface of bulk liquid having the same relative $A - B$ composition as the droplet that is in contact with an A-B gas mixture in equilibrium with the liquid solution at temperature T. Note that the gas composition (i.e., the relative proportions of A and B in the gas phase) for equilibrium with the liquid solution is not necessarily the same as the $n - m$ composition in the droplet.

Using equation (4.4) in equation (4.2), and rearranging, gives

$$\Delta G_{nm}(n,m,p_A,p_B,T) = n\left[\tilde{G}_{A,l}(p_A) - \tilde{G}_{A,v}(p_A)\right] + m\left[\tilde{G}_{B,l}(p_B) - \tilde{G}_{B,v}(p_B)\right] + \sigma s_{nm}. \qquad (4.5)$$

In the case of single-component nucleation, equilibrium between a vapor and a flat surface of bulk liquid of the same substance at temperature T occurs at the substance's saturation vapor pressure, $p_s(T)$. In the case of binary nucleation, the analogous equilibrium partial pressures, denoted $p_{A,eq}$ and $p_{B,eq}$, are the vapor pressures of A and B, respectively, for equilibrium of these vapors with a flat surface of bulk liquid having the same A-B composition as the droplet, at the same temperature T. These pressures are not the same as the saturation vapor pressures $p_{s,A}$ and $p_{s,B}$, as these latter pressures refer to equilibrium of the vapor with pure liquid A and pure liquid B, respectively.

While the saturation ratio $S = p_1/p_s$ is the driving force for single-component nucleation, the analogous role in binary nucleation is played by the pressure ratios $p_A/p_{A,eq}$ and $p_B/p_{B,eq}$. These latter pressure ratios can thus be said to represent the supersaturation of each component vapor with respect to a bulk binary liquid solution having the same relative A-B composition as the cluster $A_n B_m$. Let us denote them by S'_A and S'_B, respectively:

$$S'_A \equiv \frac{p_A}{p_{A,eq}}, \qquad (4.6)$$

$$S'_B \equiv \frac{p_B}{p_{B,eq}}. \qquad (4.7)$$

Treating an individual cluster as an incompressible liquid and the monomer vapor as an ideal gas, we can relate the free energies evaluated at the actual partial pressures in the gas, p_A and p_B, to their values at $p_{A,eq}$ and $p_{B,eq}$, respectively. For the liquid free energies, from equation (2.83),

$$\tilde{G}_{A,l}(p_A) = \tilde{G}_{A,l}(p_{A,eq}) + \tilde{v}_A(p_A - p_{A,eq}), \qquad (4.8)$$

and

$$\tilde{G}_{\text{B},l}(p_\text{B}) = \tilde{G}_{\text{B},l}(p_{\text{B},eq}) + \tilde{v}_\text{B}(p_\text{B} - p_{\text{B},eq}),\tag{4.9}$$

where $\tilde{v}_\text{A}$ and $\tilde{v}_\text{B}$ are the molecular volumes of A and B, respectively, while for the gas-phase free energies, from equation (2.76),

$$\tilde{G}_{\text{A},v}(p_\text{A}) = \tilde{G}_{\text{A},v}(p_{\text{A},eq}) + k_\text{B}T\ln\frac{p_\text{A}}{p_{\text{A},eq}},\tag{4.10}$$

and

$$\tilde{G}_{\text{B},v}(p_\text{B}) = \tilde{G}_{\text{B},v}(p_{\text{B},eq}) + k_\text{B}T\ln\frac{p_\text{B}}{p_{\text{B},eq}}.\tag{4.11}$$

Inserting these relations into equation (4.5), and rearranging, gives

$$\Delta G_{nm}(n,m,p_\text{A},p_\text{B},T) = n\left[\tilde{G}_{\text{A},l}(p_{\text{A},eq}) - \tilde{G}_{\text{A},v}(p_{\text{A},eq}) + \tilde{v}_\text{A}(p_\text{A} - p_{\text{A},eq}) - k_B T\ln\left(\frac{p_\text{A}}{p_{\text{A},eq}}\right)\right]$$
$$+ m\left[\tilde{G}_{\text{B},l}(p_{\text{B},eq}) - \tilde{G}_{\text{B},v}(p_{\text{B},eq}) + \tilde{v}_\text{B}(p_\text{B} - p_{\text{B},eq})\right.$$
$$\left. - k_B T\ln\left(\frac{p_\text{B}}{p_{\text{B},eq}}\right)\right] + \sigma s_{nm}.$$

$$\tag{4.12}$$

As in the development of CNT for single-component nucleation, we note that for ideal gases as well as for ideal liquid solutions, Gibbs free energy per molecule is identical to the molecular chemical potential. Thus equation (4.12) can be rewritten as

$$\Delta G_{nm}(n,m,p_\text{A},p_\text{B},T) = n\left[\tilde{\mu}_{\text{A},l}(p_{\text{A},eq}) - \tilde{\mu}_{\text{A},v}(p_{\text{A},eq}) + \tilde{v}_\text{A}(p_\text{A} - p_{\text{A},eq}) - k_B T\ln\left(\frac{p_\text{A}}{p_{\text{A},eq}}\right)\right]$$
$$+ m\left[\tilde{\mu}_{\text{B},l}(p_{\text{B},eq}) - \tilde{\mu}_{\text{B},v}(p_{\text{B},eq}) + \tilde{v}_\text{B}(p_\text{B} - p_{\text{B},eq}) - k_B T\ln\left(\frac{p_\text{B}}{p_{\text{B},eq}}\right)\right] + \sigma s_{nm}.$$

$$\tag{4.13}$$

Then, noting that $p_{\text{A},eq}$ and $p_{\text{B},eq}$ are by definition the equilibrium vapor pressures of substances A and B above a flat surface of bulk liquid having the same composition as the cluster $A_n B_m$, and invoking the condition for equilibrium between two or more phases of the same substance, namely that their molecular chemical potentials must be equal, gives

$$\Delta G_{nm}(n,m,p_\text{A},p_\text{B},T) = n\left[\tilde{v}_\text{A}(p_\text{A} - p_{\text{A},eq}) - k_B T\ln\left(\frac{p_\text{A}}{p_{\text{A},eq}}\right)\right]$$
$$+ m\left[\tilde{v}_\text{B}(p_\text{B} - p_{\text{B},eq}) - k_B T\ln\left(\frac{p_\text{B}}{p_{\text{B},eq}}\right)\right] + \sigma s_{nm}.$$

$$\tag{4.14}$$

Finally, as in equation (3.21), we note that the first term inside each of the brackets on the right-hand side of equation (4.14) is usually much smaller than the second term and can therefore, in most cases, be neglected. With this approximation, we now have

$$\Delta G_{nm} = -nk_BT \ln\left(\frac{p_A}{p_{A,eq}}\right) - mk_BT \ln\left(\frac{p_B}{p_{B,eq}}\right) + \sigma s_{nm}. \qquad (4.15)$$

It can be noticed that this expression seemingly suffers from the same lack of self-consistency as the standard CNT model for single-component nucleation, discussed in Section 3.10, as neither $\Delta G_{1,0}$ nor $\Delta G_{0,1}$ – the free energy of formation of a single molecule of either A or B – as given by equation (4.15) equals zero. Various modifications have been proposed for making equation (4.15) self-consistent (Flagan 2007; Kulmala et al. 1992; Wilemski and Wyslouzil 1995). However, it might be argued that equation (4.15) should apply only to the case where *both* n and m are nonzero, as otherwise the cluster would have only a single component, and a self-consistent expression such as equation (3.25) can then apply. Moreover, as the theory assumes that the smallest binary compound, A_1B_1, is formed by the nontrivial reaction

$$A_1 + B_1 \rightarrow A_1B_1, \qquad (4.16)$$

there is no requirement that $\Delta G_{1,1}$ must equal zero.

In the single-substance case – that is, where either n or m equals zero – equation (4.15) is seen to be identical to equation (3.104) in the standard (non-self-consistent) CNT model, as in that case the pressure ratios in equation (4.15) are equal to the single-substance saturation ratios.

The pressure ratios in equation (4.15) are often recast in terms of the activities of each substance in each phase. The gas-phase activity $a_{A,v}$ of substance A is defined by

$$a_{A,v} \equiv \frac{p_A}{p_{s,A}}. \qquad (4.17)$$

Thus $a_{A,v}$ is identical to the saturation ratio, S_A, of substance A, with respect to a flat surface of pure liquid A.

The liquid-phase activity $a_{A,l}$ of substance A is defined in relation to a multicomponent liquid of given composition. With reference to bulk liquid having the same composition as cluster A_nB_m,

$$a_{A,l}(n,m,T) \equiv \frac{p_{A,eq}(n,m,T)}{p_{s,A}(T)}. \qquad (4.18)$$

With these definitions, equation (4.15) can alternatively be written as

$$\Delta G_{nm} = -nk_BT \ln\frac{a_{A,v}}{a_{A,l}} - mk_BT \ln\frac{a_{B,v}}{a_{B,l}} + \sigma S_{nm}. \qquad (4.19)$$

At given temperature, the liquid-phase activities as well as surface tension are material properties that depend on droplet composition, while the vapor-phase activities depend on the values of p_A and p_B, or, equivalently, the pure-substance saturation ratios.

With these considerations, one can construct a table or graph of $\Delta G(n,m)$ for a given pair of substances, for given values of their gas-phase activities and temperature. This graph, which requires knowledge of the composition-dependent liquid-phase activities

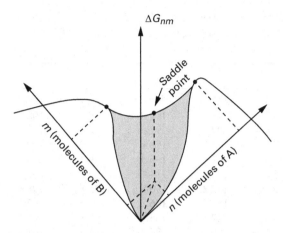

Figure 4.2 Example of free energy of formation, $\Delta G_{nm}(n, m; p_A, p_B, T)$ of a binary cluster containing n molecules of substance A and m molecules of substance B.

and surface tensions for all possible values of n and m, up to the largest sizes considered, is a three-dimensional surface that typically has multiple local maxima and minima. More generally, for multicomponent nucleation that involves droplets consisting of N substances, the graph of ΔG is an $(N + 1)$-dimensional surface, with each axis representing the number of molecules of one of the substances in the droplet. The ΔG curve along each of these axes, which represents the pure-substance ΔG_n of CNT, has a maximum, assuming that the saturation ratio for that substance lies within the condensation/evaporation regime discussed in Section 3.5.

For clusters containing multiple components, the ΔG surface exhibits saddle points, at which the slope in each orthogonal direction equals zero. For binary nucleation, there is a single saddle point, at which the surface curves upward on both sides of the saddle point in one direction and curves downward on both sides in the orthogonal direction, as with a mountain pass, as illustrated in Figure 4.2.

In the former direction, the saddle point is located at a relative minimum between two peaks. In the latter direction, which would be the crossing direction of a mountain pass, the saddle point is located at a relative maximum. In the case of a mountain pass, the saddle point is at a relatively low elevation, compared to the upward slopes on either side, and thus represents the lowest elevation for crossing over the mountain. Likewise, the ΔG saddle point represents the lowest free energy barrier to nucleation, on a landscape with multiple free energy barriers. Thus a cluster growth trajectory that crosses the saddle point is thermodynamically the most favorable cluster growth path, and the cluster size and composition at the saddle point constitute the critical cluster size and composition (n^*, m^*).

To find the values of n^* and m^*, one can treat the discrete number n and m of A and B monomers in the cluster as continuous variables x and y, respectively, so that equation (4.15) can be differentiated in each direction. At the saddle point, *both*

$$\left(\frac{\partial \Delta G}{\partial x}\right)_y = 0, \tag{4.20}$$

Table 4.1 Example calculation of dimensionless free energies of cluster formation, $\Delta G_{nm}/k_B T$, near the saddle point, for $(H_2O)_n(H_2SO_4)_m$ clusters under conditions typical of the stratosphere: $T = 218$ K; $p_{H_2O} = 0.0267$ Pa; $p_{H_2SO_4} = 1.53 \times 10^{-9}$ Pa.

n\m	75	76	77	78	79	80	81	82
128	187.5	193.5	197.5	199.5	200.1	199.1	197.5	194.8
129	183.2	190.5	195.5	198.5	199.8	200.1	198.8	196.5
130	187.2	186.5	192.8	197.1	199.5	200.1	199.8	198.1
131	190.5	184.2	189.5	194.8	198.1	199.8	200.1	199.5
132	196.5	187.8	**184.8**	191.8	196.5	199.1	200.1	200.1
133	195.8	191.2	185.2	188.2	194.1	197.8	199.8	202.1
134	197.8	193.8	188.8	183.2	190.8	195.8	198.8	200.1
135	199.5	196.5	191.8	186.2	186.8	193.2	197.1	199.5
136	200.8	198.5	194.5	189.5	183.2	189.8	195.1	198.5
137	201.5	199.8	196.8	192.5	187.2	185.2	192.2	196.5
138	202.1	201.1	198.8	195.1	190.2	184.2	188.5	194.1
139	201.8	201.8	200.1	197.5	193.2	187.5	183.5	190.8

Adapted from Hamill et al. (1977).
Shaded cells indicate the minimum value ΔG_{nm} for each value of m. Among these minima, the maximum value of ΔG_{nm} determines the location of the saddle point, indicated in bold.

and

$$\left(\frac{\partial \Delta G}{\partial y}\right)_x = 0 \tag{4.21}$$

must apply, where

$$\Delta G_{xy} = -x k_B T \ln\left(\frac{p_A}{p_{A,eq}}\right) - y k_B T \ln\left(\frac{p_B}{p_{B,eq}}\right) + \sigma s_{xy}. \tag{4.22}$$

A difficulty in carrying out these differentiations is that the equilibrium partial pressures (equivalently, the liquid-phase activities) as well as surface tension depend not only on temperature but also on cluster composition. Of course, if one knows the values of these dependencies for all values of (n, m), up to some suitable size, then one can construct a table of $\Delta G(n, m)$ for given values of the gas-phase partial pressures and temperature. For example, Table 4.1 shows an adapted version of a table of free energies of formation from calculations of Hamill et al. (1977), for a mixture of sulfuric acid and water, and for values of the temperature and partial pressures of these substances that are representative of conditions in the earth's stratosphere. For these conditions, these calculations indicate that the saddle point in the free energy landscape occurs for a cluster consisting of 132 molecules of H_2O and 77 molecules of H_2SO_4.

However, lacking such property information, classical theory makes the assumption that the dominant cluster growth path crosses the saddle point in a direction that maintains constant composition, that is, constant proportions of A and B. As cluster growth involves discrete events in which a single molecule of either A or B adds to the cluster, this is obviously a poor approximation at small cluster sizes. Arguably it becomes a better approximation as clusters become large, which is consistent with the spirit of CNT.

Assuming then, as in single-component CNT, that the droplet's surface tension obeys the capillarity approximation, that is, that it is the same as for a flat surface of bulk liquid having the same composition as the droplet, in equilibrium with its two-component vapor, and that cluster growth follows a trajectory in the vicinity of the saddle point that maintains constant composition, the equilibrium partial pressures and surface tension can be treated as constant. In that case, differentiating equation (4.22) with respect to x gives

$$\left(\frac{\partial \Delta G_{xy}}{\partial x}\right)_y = -k_B T \ln\left(\frac{p_A}{p_{A,eq}}\right) + \sigma\left(\frac{\partial s_{xy}}{\partial x}\right)_y. \tag{4.23}$$

Converting from surface area to radius for a spherical droplet and making the approximation that the droplet volume equals the sum of the molecular volumes of the x A molecules and y B molecules comprising the droplet:

$$v_{xy} \approx x\tilde{v}_A + y\tilde{v}_B, \tag{4.24}$$

one obtains

$$\left(\frac{\partial s_{xy}}{\partial x}\right)_y = \frac{2\tilde{v}_A}{r_{xy}}. \tag{4.25}$$

Using this result in equation (4.23), and satisfying the condition (4.20), one obtains the following result for the radius r^* of the critical cluster:

$$r^* = \frac{2\sigma \tilde{v}_A}{k_B T \ln(p_A/p_{A,eq})}. \tag{4.26}$$

Following the same derivation, an analogous relation must also apply for substance B, as well as for any substance in a multicomponent critical cluster:

$$r^* = \frac{2\sigma \tilde{v}_B}{k_B T \ln(p_B/p_{B,eq})}. \tag{4.27}$$

Thus the right-hand sides of equations (4.26) and (4.27) are equal. Rearranging, the critical cluster composition must satisfy

$$\left[\frac{\ln(p_B/p_{B,eq})}{\ln(p_A/p_{A,eq})}\right]^* = \frac{\tilde{v}_B}{\tilde{v}_A}, \tag{4.28}$$

or, using the effective supersaturations defined by equations (4.6) and (4.7),

$$\left(\frac{\ln S'_B}{\ln S'_A}\right)^* = \frac{\tilde{v}_B}{\tilde{v}_A}. \tag{4.29}$$

Thus, according to classical theory, the effective supersaturation of each component with respect to the composition of the critical cluster bears a remarkably simple relationship to the molecular volume of each component. The source of this perhaps surprising result lies in the assumption that surface tension in the vicinity of the critical size is constant, together with equation (4.25), which is a simple consequence of geometry.

Considering equation (4.28), it can be noted that p_A and p_B are the actual partial pressures of components A and B. On the other hand, $p_{A,eq}$ and $p_{B,eq}$ depend on the cluster composition. Therefore, the relative proportions of A and B in the critical cluster must be such that equation (4.28) is satisfied.

As $p_{A,eq}$ and $p_{B,eq}$ are the equilibrium vapor pressures above a flat surface of bulk liquid having the same composition, they depend on the mole fractions of each substance in the cluster but not on the absolute number of molecules of each species in the cluster:

$$p_{A,eq} = p_{A,eq}(\chi_A, T),$$
(4.30)

and

$$p_{B,eq} = p_{B,eq}(\chi_B, T),$$
(4.31)

where χ_A and χ_B are the mole fractions of A and B, respectively, in the cluster.

However, for a two-component cluster, these mole fractions are not independent, as

$$\chi_A + \chi_B = 1.$$
(4.32)

Equations (4.26), (4.27), and (4.30)–(4.32) can be regarded as constituting five equations for the five unknown variables, χ_A, χ_B, $p_{A,eq}$, $p_{B,eq}$, and r^*. Solving this, the values of n^* and m^* can then be determined from the values of either of the mole fractions together with the size of the cluster, r^*. The value of ΔG^* then follows from equation (4.15).

Combining equations (4.15), (4.24), and (4.26), a simple relationship is obtained between the surface energy of the critical cluster and its free energy of formation:

$$\Delta G^* = \frac{\sigma s^*}{3}.$$
(4.33)

4.3   Nucleation Involving an Arbitrary Number of Components

The extension of CNT for binary nucleation to the more general case of multicomponent nucleation involving an arbitrary number of components is in principle straightforward.

Consider a gas containing condensible vapors of an arbitrary number of species A, B, C, ... that form clusters containing n_i monomers of each species i. Making the same assumptions as in Section 4.2 for binary CNT, and following the same derivation, the free energy of cluster formation can be written as

$$\Delta G_{n_A, n_B, n_C, ...} = \sigma s - k_B T \sum_i n_i \ln\left(\frac{p_i}{p_{i,eq}}\right),$$
(4.34)

or, in terms of activities,

$$\Delta G_{n_A,n_B,n_C,\ldots} = \sigma s - k_B T \sum_i n_i \ln\left(\frac{a_{i,v}}{a_{i,l}}\right). \tag{4.35}$$

Each species i of the critical cluster satisfies the relation

$$r^* = \frac{2\sigma \tilde{v}_i}{k_B T \ln\left(p_i/p_{i,eq}\right)}. \tag{4.36}$$

The mole fractions of the species comprising a cluster must add to unity,

$$\sum_i \chi_i = 1, \tag{4.37}$$

and, as all pairs of species must satisfy equation (4.29), it follows that the composition of the critical cluster must be such that the effective supersaturations of each species are in proportion to their molecular volumes:

$$\ln\left(S_i'\right)^* \propto \tilde{v}_i. \tag{4.38}$$

Finally, equation (4.33) applies, regardless of the number of components in the critical cluster.

In binary nucleation, the ΔG surface is three-dimensional, with the critical cluster size and composition located at a saddle point, where ΔG is a maximum in one direction and a minimum in the orthogonal direction. In general, if there are M components in the cluster, then the ΔG surface is $(M+1)$-dimensional, and the critical cluster is located at the point where ΔG is a maximum in one direction and a minimum in all orthogonal directions.

Regardless of the straightforwardness of the extension from binary nucleation to the case of more than two components, the caveats noted in Section 4.2 regarding the difficulty in locating the critical cluster on the ΔG surface – particularly regarding the dependence of both surface tension and equilibrium vapor partial pressures on cluster composition – become even more problematic as the number of components increases.

4.4 Advances within the Classical Framework

The classical model for multicomponent nucleation presented above retains the same key simplifications as in the single-component CNT. In doing so, it ignores several complications that multicomponent nucleation presents, which are not encountered in the single-component case. A number of studies have attempted to develop improved models that take these difficulties into account, while still retaining the essence of the liquid droplet model.

Among these difficulties are the fact that many different reaction sequences that grow a cluster to critical size and composition are possible; the fact that, even within the capillarity approximation, surface tension depends on cluster composition; and the fact that the composition at the surface of a droplet may differ from that in the core.

4.4.1   The Cluster Growth Trajectory

In the theory of single-component nucleation, there is only one sequence of reactions that grows a cluster from monomer A to cluster A_n, namely that given by reactions (2.6)–(2.9). However, in binary nucleation, there are many possible reaction sequences that could grow a cluster from A and B monomers to size A_nB_m, even assuming that the overall reaction can be written as in reaction (4.1). As cluster growth occurs by discrete monomer addition events, the actual growth path must inevitably be some sort of zigzag, with many such zigzag paths possible.

The classical theory of binary nucleation assumes that there exists a dominant cluster growth path that passes through the saddle point on the free energy landscape and that the nucleation rate is strongly dominated by the "flow" of clusters, analogous to the nucleation current of single-component nucleation theory, in the vicinity of the saddle point. Based on purely thermodynamic considerations, Reiss (1950) assumed that the growth trajectory would follow the path of steepest ascent to, and descent from, the saddle point. However, a complication is that the proportions of the two substances are different in the gas phase than in the cluster, implying that the collision rates of the two vapor species with the cluster are likely in different proportion than the cluster composition. Thus, if one uses equation (3.94) to calculate the steady-state nucleation rate,

$$J = Zf^* \times (N_{n^*})_{eq}, \tag{3.94}$$

where f^* is the rate (s^{-1}) of monomer addition to a single critical cluster, one must suitably average the separate arrival rates of monomers of species A and B, and this requires knowledge of the flow path, which in general is not the same as the path of steepest descent from the saddle point, as that depends only on thermodynamic considerations.

Considering this, Stauffer (1976) developed a theory for the dominant flow path that combines thermodynamic and kinetic factors. For a detailed derivation of the critical cluster flow vector (magnitude and direction) in steady-state multicomponent nucleation, see Vehkamäki (2006).

4.4.2   Composition-Dependent Surface Tension

In differentiating ΔG_{xy} with respect to both x and y so as to locate the saddle point in the free energy landscape, equation (4.23) assumes that the surface tension can be treated as constant in the vicinity of the saddle point. This assumption is problematic. Even if the relative proportions of A and B can be treated as constant along the dominant cluster flow path, the growth trajectory is not in general parallel to either the x or y composition axes. Thus, as surface tension is generally a sensitive function of cluster composition, there is no reason to expect either $(\partial\sigma/\partial x)_y$ and $(\partial\sigma/\partial y)_x$ to equal zero at the saddle point. For the assumption of constant surface tension to be reasonable, then either the cluster should be very large or one of the components in the critical cluster, in the case of binary clusters, should be quite dilute with respect to the other.

In Table 4.1, the critical cluster in binary water-sulfuric acid nucleation is identified, for the conditions of that table, as $(H_2O)_{132}(H_2SO_4)_{77}$, which is arguably large enough for the assumption of constant surface tension to be reasonable. On the other hand, one can expect critical sizes in many scenarios to be much smaller. For example, in a CNT-based study of ternary nucleation in the sulfuric acid-ammonia-water system, Napari et al. (2002) found the critical cluster, for the conditions they examined, to consist almost entirely of sulfuric acid and ammonia and to contain fewer than 10 molecules. At that size, the addition of one molecule to the cluster could obviously have a major effect on surface tension, even under the capillarity approximation.

Regarding the hypothetical situation where one of the components in a binary cluster may be quite dilute with respect to the other, it is interesting to note that in the most studied binary nucleation system, sulfuric acid and water, sulfuric acid in the vapor phase is known to promote nucleation even at extremely low concentrations. Nevertheless, in general the mole fraction of sulfuric acid in the critical cluster is not necessarily small.

If constant surface tension near the saddle point is not a reasonable assumption, then, instead of equation (4.23), one has

$$\left(\frac{\partial \Delta G_{xy}}{\partial x}\right)_y = -k_B T \ln\left(\frac{p_A}{p_{A,eq}}\right) + \sigma\left(\frac{\partial s_{xy}}{\partial x}\right)_y + s_{xy}\left(\frac{\partial \sigma}{\partial x}\right)_y, \tag{4.39}$$

with an analogous expression for differentiation with respect to y. The last term, involving the gradients of composition-dependent surface tension in the x- and y-directions, affects the location of the saddle point and hence the value of ΔG^*, which thus may differ from the case of constant surface tension.

4.4.3   Effect of Surface Active Layer

In Section 3.7.1, we discuss the existence of a thin transition layer at the surface of a droplet in single-component nucleation and its effect on surface tension. In multicomponent nucleation, some components of the nucleating droplets may be surface active, meaning that these components are more concentrated in the surface monolayer than in the core. An example is binary nucleation of water with various alcohols, where the alcohol typically concentrates near the surface, lowering the surface tension. Several authors have developed models that account for this effect within the context of CNT for binary nucleation (Flageollet-Daniel et al. 1983; Laaksonen 1992; Mirabel and Katz 1974).

Homework Problems

4.1.   Consider the formation of $A_n B_m$ clusters by sequential reactions with A and B monomers, in any order. In general, how many reactions are involved in each sequence? In general, how many possible such sequences exist?

4.2.   Considering the fact that the surface tension σ of a droplet may depend on the droplet's composition, does the specific sequence of reactions leading to formation of cluster $A_n B_m$ affect the value of the cluster's free energy of formation, ΔG_{nm}? Verify your answer by considering the possible reaction sequences involved in forming A_2B_2 and examining the stepwise reactions associated with each sequence.

4.3 Derive equation (4.25).

4.4 Consider ternary nucleation involving clusters with components A, B, and C, denoted $A_nB_mC_l$. Develop a set of relations, analogous to the discussion around equations (4.26)–(4.33), that would allow one to determine the values of n^*, m^*, and l^* in the critical cluster.

4.5 Derive equation (4.33).

5 Classical Theory of Ion-Induced Nucleation

J. J. Thomson, who was awarded the 1906 Nobel Prize in Physics for studies of the conduction of electricity in gases, observed in his 1888 text, *Applications of Dynamics to Physics and Chemistry*, that electrically charging a liquid caused the vapor pressure above the liquid to decrease (Thomson 1888). Thomson reasoned that the charge of the liquid exerts an attractive force on vapor molecules that promotes their condensation. Considering this effect in the situation where a gas includes condensible vapor as well as dispersed, charged droplets, the droplets may act as condensation nuclei. If droplets may form initially by condensation around an ion, it follows that the presence of ions can cause nucleation to occur at lower saturation ratios than would be the case in the absence of ions.

Here we discuss the classical theory of ion-induced nucleation, based on Thomson's ideas about ionic clusters. Typically ion-induced nucleation is treated as a type of heterogeneous nucleation, in contrast to the homogeneous nucleation scenarios discussed to this point. However, from the viewpoint of chemical kinetics, reactions between ions and neutral molecules, where both are in the gas phase, are considered homogeneous reactions, and the term "heterogeneous" usually implies reactions between gas-phase species and solid or liquid surfaces. We adopt that approach here and thus include ion-induced nucleation among the types of gas-phase nucleation that we consider.

In this chapter, we discuss the effect of ions on nucleation under nonplasma conditions – that is, under conditions where the existence of free electrons and electric fields is relatively unimportant. The case of nucleation in plasmas is considered in Chapter 9.

5.1 The Wilson Cloud Chamber

In the latter years of the nineteenth century, C. T. R. Wilson, who had been Thomson's student, invented a device called the "cloud chamber," with which he conducted experiments (Wilson 1897, 1899) that led to his being awarded the 1927 Nobel Prize in Physics, "for his method of making the paths of electrically charged particles visible by condensation of vapour" (NobelPrize.org 2021). The cloud chamber played a key role in early-twentieth-century studies of charged particles and X-rays. The physicist Ernest Rutherford, himself winner of the 1908 Nobel Prize in Chemistry, remarked that the cloud chamber was "the most original and wonderful instrument in scientific history" (Nobel Foundation 1965).

While Wilson was awarded the Nobel Prize chiefly for his visualization of the paths of ions and electrons, that was not his initial motivation for developing the cloud chamber. Wilson was primarily a meteorologist and thus was interested in condensation of water vapor. The cloud chamber was designed to produce rapid expansions, which, as discussed in Section 3.11.1, can generate supersaturated vapors, leading to nucleation. Wilson conducted experiments in which he produced rapid volumetric expansions with various expansion ratios. If an expansion is sufficiently rapid, it can be assumed to be adiabatic, as heat loss to the walls occurs on timescales that are much longer than the expansion time. If the piston motion is further idealized as being frictionless, hence reversible, then the expansion is isentropic, and one can assume the simple relationships among changes in volume, temperature, and pressure presented in equations (3.114) and (3.115). Knowing the initial conditions of the air or other carrier gas, one can then infer the vapor saturation ratio at the end of the expansion.

For his conditions, Wilson found that there existed a critical volumetric expansion ratio of about 1.25, which corresponded to a vapor saturation ratio of approximately four (Wilson 1927). Above this critical expansion ratio, a shower of drops was observed to fall. Wilson found that he could repeat the expansion over and over and keep getting the same result. He reasoned that ions were continually generated in the chamber by ionizing radiation, in the form of X-rays, and that these ions served as condensation nuclei. When water condensed onto these nuclei, the droplets grew until they were large enough to fall by gravity, so that the ions were removed from the chamber, until a new generation of ions was generated, repeating the process.

When Wilson improved his apparatus to allow for more rapid expansions, he was able to achieve higher supersaturations before condensation was observed. He found that a second critical saturation ratio existed for his conditions, at a value of approximately eight, above which dense clouds were formed. For saturation ratios between these two critical values, the number of drops did not vary, consistent with the hypothesis that ions were naturally generated at a constant rate, and served as condensation nuclei. For saturation ratios above the second critical value, the number of drops increased rapidly with increasing supersaturation. Wilson reasoned that in this case the vapor molecules themselves acted as condensation nuclei.

In effect, the Wilson cloud chamber was the first apparatus used to study particle nucleation from the gas phase, both by ion-induced nucleation and neutral self-nucleation.

5.2   Ion-Induced Cluster Growth

We consider a scenario that is analogous to the single-component clustering sequence described in Section 2.1. In this case, however, instead of beginning the sequence with the reaction of a monomer with another monomer to form a dimer, we consider a sequence that is initiated by reaction of a neutral monomer with an ion.

Let $I \cdot A_n$ stand for an ionic cluster, consisting of an ion, I, plus n monomers of the condensible vapor. The ion may be either positively or negatively charged and is not necessarily of the same chemical species as the condensible vapor. In the special case where the ion is of the same substance as the condensing vapor, $I \cdot A_n$ is equivalent to $A_{n+1}^{\pm z}$, where z represents the number of charges on the ion.

The sequence of reversible condensation/evaporation reactions involved in forming an ionic n-mer can be written as follows:

$$I + A_1 \rightleftarrows I{\cdot}A \tag{5.1}$$

$$I{\cdot}A_1 + A_1 \rightleftarrows I{\cdot}A \tag{5.2}$$

$$I{\cdot}A_2 + A_1 \rightleftarrows I{\cdot}A \tag{5.3}$$

$$\vdots$$

$$I{\cdot}A_{n-1} + A_1 \rightleftarrows I{\cdot}A_n \tag{5.4}$$

The sequence as written assumes that the initiating reaction involves the reaction of an ion with a monomer and thus neglects reactions such as

$$I + A_2 \rightleftarrows I{\cdot}A_2, \tag{5.5}$$

which could then be followed by reaction (5.3) and so forth. However, under the assumption that neutral monomers are much more abundant than neutral dimers, trimers, etc., the reaction given by equation (5.1) is much more likely than that given by (5.5), and thus the sequence (5.1)–(5.4) is the most likely ion-induced clustering path.

This does not rule out the neutral self-nucleation path described in Chapter 2, which may occur in parallel with ionic clustering. Whether ion-induced nucleation or self-nucleation dominates for a given chemical system depends on the relative abundances of ions and neutral vapor monomers. However, it is known that ion-induced nucleation may represent the dominant path even when ions are far less abundant than the vapor monomers. The reasons for this lie in both kinetics and thermodynamics – in kinetics, because ion-neutral reactions are typically faster than neutral-neutral reactions; in thermodynamics, because the stepwise Gibbs free energy changes associated with the monomer addition reactions, (5.1)–(5.4), may be smaller than in the corresponding neutral clustering mechanism, reactions (2.6)–(2.9).

5.3   The Thomson Model of an Ionic Cluster

Thomson developed an expression for the effect of charge on a droplet on the equilibrium vapor pressure around the droplet – in effect, an additional term to be included in the Kelvin equation, equation (3.35). This expression was further developed by Tohmfor and Volmer, who, in 1938, using a similar approach as presented in Chapter 3 for classical nucleation theory (CNT), developed an expression for the steady-state rate of ion-induced nucleation (Tohmfor and Volmer 1938; Volmer 1939).

5.3.1   The Classical Model for the Free Energy of Formation of an Ionic Cluster

Thomson's model is the liquid droplet model of CNT, modified such that the Gibbs free energy change associated with forming a cluster accounts for the presence of an assumed spherical ionic core of radius r_i and charge q, around which vapor condenses,

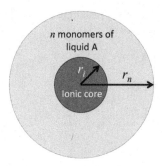

Figure 5.1 Thomson model of droplet with ionic core.

forming a spherical droplet of radius r_n, as illustrated in Figure 5.1. The ion is assumed to lie at the center of this droplet.

An ion induces an electric dipole in a nearby neutral molecule, causing the ion and the neutral molecule to experience an attractive potential relative to each other. Evaporation of a monomer from an ionic droplet requires work against this attractive potential. Let us denote this work by $W_{E,1}$, the subscript "E" indicating that the work is done against the electric force and the subscript "1" denoting evaporation of one monomer. The work done against the attractive potential associated with *adding* a monomer is the negative of this term.

Following the discussion in Section 3.3.3, and letting $\Delta G_{n-1,n}$ now stand for the stepwise free energy change associated with reaction (5.4), the expression for $\Delta G_{n-1,n}(p_s)$ under the capillarity approximation is now a modified version of that given by equation (3.23). In that equation, $\Delta G_{n-1,n}(p_s)$ is given simply by the change in surface energy associated with addition of one monomer. For an ionic monomer, Thomson additionally accounted for the work done against the attractive potential and assumed that the surface tension term was independent of the electrical work term. Thus

$$\Delta G_{n-1,n}(p_s) = \sigma\,(s_n - s_{n-1}) - W_{E,1}, \tag{5.6}$$

where s_n stands for the surface area of the ionic droplet $I \cdot A_n$.

The overall reaction to form an ionic cluster from condensation of n monomers onto an ion is given by

$$I + nA_1 \rightarrow I \cdot A_n. \tag{5.7}$$

As given by equation (2.26), the free energy change $\Delta G_n(p_s)$ for this reaction can be obtained by summing over all of the stepwise reactions in the sequence (5.1)–(5.4). In doing this, the first reaction, (5.1), requires special attention, as the ionic cluster on the left-hand side is simply the bare ion itself, which, following our notation, could be written as $I \cdot 0$.

Let $\Delta G_{0,1}$ denote the stepwise free energy change for reaction (5.1), and let us assume that the electrical work associated with evaporation of one monomer is independent of cluster size. Then the stepwise free energy changes for the reactions in the sequence (5.1)–(5.4) are as follows:

$$I \cdot 0 + A_1 \rightleftarrows I \cdot A_1, \Delta G_{0,1}(p_s) = \sigma(s_1 - s_i) - W_{E,1} \tag{5.8}$$

$$I \cdot A_1 + A_1 \rightleftarrows I \cdot A_2, \Delta G_{1,2}(p_s) = \sigma(s_2 - s_1) - W_{E,1} \tag{5.9}$$

$$I \cdot A_2 + A_1 \rightleftarrows I \cdot A_3, \Delta G_{2,3}(p_s) = \sigma(s_3 - s_2) - W_{E,1} \qquad (5.10)$$

$$\vdots$$

$$I \cdot A_{n-1} + A_1 \rightleftarrows I \cdot A_n, \Delta G_{n-1,n}(p_s) = \sigma(s_n - s_{n-1}) - W_{E,1} \qquad (5.11)$$

Overall reaction: $I + nA_1 \rightarrow I \cdot A_n, \Delta G_n(p_s) = \sigma(s_n - s_i) - nW_{E,1}$ \qquad (5.12)

Note that, in contrast to equation (3.25), the surface area of the ion, rather than that of the monomer, is subtracted from s_n, as the first reaction in the sequence involves condensation onto an ion rather than onto a monomer. Thomson, in his model, neglected to subtract the surface area of the ion. Thus his surface energy term in ΔG_n is the same as in the standard version of CNT for neutral self-nucleation.

Applying equation (2.77) to equation (5.12), the value of ΔG_n evaluated at actual monomer partial pressure p_1 is given by

$$\Delta G_n(p_1) = \sigma(s_n - s_i) - W_{E,n} - nk_B T \ln S, \qquad (5.13)$$

where we have utilized

$$W_{E,n} = nW_{E,1}. \qquad (5.14)$$

Note that the last term on the right-hand side of equation (5.13) differs from the corresponding term in the case of single-component self-nucleation, equation (3.25), which is given by $(n - 1)k_B T \ln S$. The reason for this difference is simply that formation of $I \cdot A_n$ via sequence (5.1)–(5.4) requires n stepwise reactions, whereas forming A_n by neutral clustering requires $(n - 1)$ stepwise reactions.

5.3.2 Work Done by or against the Attractive Potential

$W_{E,n}$ represents the work done against the attractive potential required to remove n monomers from the ionic cluster. To evaluate this term, Thomson modeled the ionic cluster as an electric capacitor.

The capacitance C of a cluster can be written as

$$C = \varepsilon C_0, \qquad (5.15)$$

where ε is the relative permittivity (dielectric constant) of the substance (assumed liquid) of the droplet and C_0 is what the capacitance would equal if the same volume were filled instead by vacuum. Relative permittivity is a material property whose value is greater than unity. For example, for water at room temperature, $\varepsilon \approx 80$.

Let ϕ denote the difference in electric potential between the surface of the droplet and the surface of the ionic core. Then, by definition of capacitance, ϕ is related to the charge on the ion and the capacitance of the droplet by

$$\phi = \frac{q}{C}. \qquad (5.16)$$

The work $W_{E,n}$ required to replace the condensed liquid with vacuum can be written as

$$W_{E,n} = U_0 - U, \tag{5.17}$$

where U is the electrical energy stored in the capacitor, given by

$$U = \frac{1}{2}C\phi^2, \tag{5.18}$$

and subscript "0" again denotes vacuum. Thus

$$U_0 = \frac{1}{2}C_0\phi_0{}^2. \tag{5.19}$$

Comparing equations (5.15)–(5.19) gives

$$W_{E,n} = U_0\left(1 - \varepsilon^{-1}\right). \tag{5.20}$$

Now, the electric field E_0 around a point charge q in vacuum is given by

$$E_0 = \frac{q}{4\pi\varepsilon_0 r^2}, \tag{5.21}$$

where ε_0 is the vacuum permittivity (8.85×10^{-12}F/m) and r is the distance from the point charge.[1] The potential difference between the surface of the droplet and the surface of the ionic core, for the vacuum case, is then obtained by integrating the electric field over this region:

$$\phi_0 = \int_{r_i}^{r_n} E_0 dr = \frac{q}{4\pi\varepsilon_0}\left(r_i^{-1} - r_n^{-1}\right). \tag{5.22}$$

From equations (5.16) and (5.19),

$$U_0 = \frac{1}{2}q\phi_0. \tag{5.23}$$

Combining this with equations (5.20) and (5.22) then gives

$$W_{E,n} = \frac{q^2}{8\pi\varepsilon_0}\left(1 - \varepsilon^{-1}\right)\left(r_i^{-1} - r_n^{-1}\right). \tag{5.24}$$

Finally, inserting this into equation (5.13), one obtains an expression for the free energy change associated with forming an ionic cluster from an ion plus n vapor-phase monomers,

$$\Delta G_n(p_1) = \sigma(s_n - s_i) - \frac{q^2}{8\pi\varepsilon_0}\left(1 - \varepsilon^{-1}\right)\left(r_i^{-1} - r_n^{-1}\right) - nk_BT \ln S. \tag{5.25}$$

This is the expression derived by Thomson, except that he neglected to subtract σs_i. Note that when $r_n = r_i$ (i.e., $n = 0$ and $s_n = s_i$), equation (5.25) yields $\Delta G = 0$, so that self-consistency is satisfied, although it is not satisfied in the original Thomson version nor in subsequent literature that uses it.

---

[1] This presentation uses SI units. Note, however, that most of the literature on the Thomson model of ion-induced nucleation uses Gaussian units, in which case the factor $4\pi\varepsilon_0$ in equation (5.21) is omitted.

Equation (5.25) can be rewritten in terms of a common cluster size dimension, for example, radius. Assuming as usual in the classical theory that clusters are spherical,

$$s_n - s_i = 4\pi\left(r_n^2 - r_i^2\right),$$ (5.26)

and

$$n = \frac{v_n - v_i}{v_m} = \frac{4\pi}{3v_m}\left(r_n^3 - r_i^3\right),$$ (5.27)

where v_m is the monomer volume.[2] With these substitutions, and treating the radius r of the charged cluster as a continuous variable, equation (5.25) becomes

$$\Delta G_r(p_1) = 4\pi\sigma\left(r^2 - r_i^2\right) - \frac{q^2(1 - \varepsilon^{-1})}{8\pi\varepsilon_0}\left(r_i^{-1} - r^{-1}\right) - \frac{4\pi kT \ln S}{3v_m}\left(r^3 - r_i^3\right).$$ (5.28)

The extrema of $\Delta G_r(p_1)$ can then be found by determining the roots of

$$\left.\frac{\partial(\Delta G_r)}{\partial r}\right|_{T,S} = 0.$$ (5.29)

Differentiating equation (5.28) then yields a fourth-order equation for the cluster radius at the extrema of $\Delta G_r(p_1)$:

$$\frac{4\pi kT \ln S}{v_m}r^4 - 8\pi\sigma r^3 + \frac{q^2(1 - \varepsilon^{-1})}{8\pi\varepsilon_0} = 0.$$ (5.30)

5.3.3 Special Case: Ion Is Same Substance (or Same Size) as Condensible Vapor

The roots of equation (5.30) are algebraically lengthy. However, an insight can be gained by making the simplifying assumption that the ions and the neutral vapor monomer consist of the same substance A. In that case, the reaction sequence, equations (5.1)–(5.4), can be rewritten, if the ions are singly charged positive, as

$$A_1^+ + A_1 \rightleftarrows A_2^+$$ (5.31)

$$A_2^+ + A_1 \rightleftarrows A_3^+$$ (5.32)

$$A_3^+ + A_1 \rightleftarrows A_4^+$$ (5.33)

$$\vdots$$

$$A_{n-1}^+ + A_1 \rightleftarrows A_n^+$$ (5.34)

If the ion is negative, or multiply charged, these equations can be rewritten accordingly. Note that in this special case the cluster size index n incorporates the ion, unlike the case in which the ion is of a different substance than the condensible vapor. In either case, n

---

[2] Note that v_1 here is the volume of the ion with one attached monomer, so $v_1 > v_m$.

represents the number of monomers in the cluster, whether neutral or charged. Thus, for the case $I = A^+$, A_n^+ is equivalent to $I \cdot A_{n-1}$, and the reaction sequence to form A_n^+ involves $(n-1)$ reactions, whereas n reactions are required to form $I \cdot A_n$.

The overall reaction to form A_n^+ is given in this special case by

$$A_1^+ + (n-1)A_1 \rightarrow A_n^+. \tag{5.35}$$

Assuming that the size of electrons is negligible compared to the size of ions or neutral monomers, the effect on cluster size of ionization or electron attachment can be neglected. Then, continuing to assume that clusters of all sizes are spherical, one can write

$$r_n = n^{1/3} r_1 \tag{5.36}$$

and

$$s_n = n^{2/3} s_1, \tag{5.37}$$

r_1 and s_1 being the radius and surface area, respectively, of the neutral monomer.

Note that if the ions are of a different substance than the neutral monomers, these relations still hold as long as the ions have approximately the same radius as the neutral monomers. In that case, the index n in equations (5.36) and (5.37) is understood to incorporate the ion and thus corresponds to $(n-1)$ monomers attached to the ion. Indeed, as radius scales on the cube root of volume, in many cases this assumption would arguably be more reasonable than the assumption that ionic clusters are spherical.

Let us define a dimensionless electrical work Φ by

$$\Phi \equiv \frac{q^2(1 - \varepsilon^{-1})}{8\pi\varepsilon_0 r_1 k_B T}. \tag{5.38}$$

For example, for water at 20°C, assuming a singly charged ion, with $r_1 = 0.192$ nm (Section 1.2.2) and $\varepsilon = 80.1$, we find $\Phi \approx 147$, while for liquid benzene, C_6H_6, with $r_1 = 0.328$ nm and $\varepsilon \approx 2.3$, $\Phi \approx 49$. On the other hand, for metals the relative permittivity is close to unity, and thus Φ is close to zero. For example, for liquid mercury, with $r_1 = 0.180$ nm and $\varepsilon \approx 1.00074$ at room temperature, one finds $\Phi \approx 0.11$.

Under these assumptions and definitions, equation (5.25) can be recast in dimensionless form as

$$\frac{\Delta G_n(p_1)}{k_B T} = \Theta(n^{2/3} - 1) - \Phi(1 - n^{-1/3}) - (n-1)\ln S. \tag{5.39}$$

This equation is seen to be identical to the analogous result for neutral clusters in the self-consistent version of CNT, equation (3.27), except that a term is added – the second term on the right-hand side – to account for the electric force that the ion exerts on the neutral monomers in the cluster. Note also that $n = 1$ here corresponds to the case of the bare ion, for which the right-hand side of equation (5.39) correctly reduces to zero.

From equation (5.39), the value of ΔG_n for ionic clusters is determined by a competition between three contributions: surface tension, the electric force exerted by the ion, and the effect of pressure (i.e., saturation ratio) on Gibbs free energy. These depend,

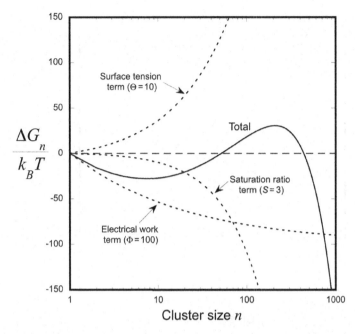

Figure 5.2 Contributions of the three terms in the Thomson model to the total Gibbs free energy of formation of ionic clusters, for the case where the ion is of the same substance as the condensible vapor, with $\Theta = 10$, $\Phi = 100$, and $S = 3$.

respectively, on the values of Θ, Φ, and S, and these three terms scale differently on cluster size. Figure 5.2 shows the contributions of the three terms to the total value of ΔG_n for the case $\Theta = 10$, $\Phi = 100$, and $S = 3$. As can be seen, the competition between these three effects results, in this case, in the existence of both a maximum and a minimum in ΔG_n.

Results for the total value of ΔG_n are shown in Figure 5.3 for the same values of Θ and Φ, but for a range of saturation ratios. Comparing the result for ΔG_n of ionic clusters to that of neutral clusters with the same value of surface tension Θ (cf. Figures 3.2 and 3.3), important qualitative differences are seen.

First, for all clusters larger than the monomer, the presence of an ion reduces ΔG_n. This is evident from equation (5.39), as relative permittivity ε is greater than unity for any real substance, so that Φ, defined by equation (5.38), is always positive for $n > 1$. This reduction in the free energy of formation due to the presence of an ion causes ΔG_n to be negative for small cluster sizes, even for saturation ratios much smaller than unity. For the dimer, one finds that ΔG_2 is negative for most values of Θ and Φ, even for saturation ratios much smaller than unity, the only exceptions being substances such as liquid mercury that have high surface tension and low relative permittivity.

Second, in Figure 5.3, one notes, for saturation ratios of 1, 3, and 4, the existence of a minimum in ΔG_n. Indeed, a minimum in ΔG_n at some size n_0 is found to exist even for saturation ratios much smaller than unity. Thus the Thomson model implies that the presence of ions leads to the existence of ionic clusters that are more thermodynamically stable than the bare ion. For example, for the conditions of Figure 5.3, and for the case

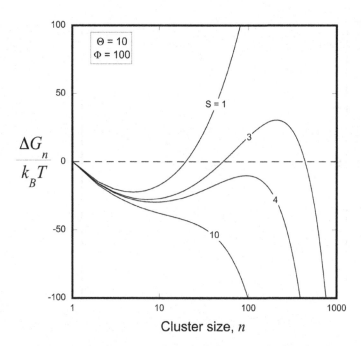

Figure 5.3 Gibbs free energy of cluster formation versus cluster size for ionic clusters, using the classical Thomson model, for the case where the ion is of the same substance as the condensible vapor, for various saturation ratios. In all cases, the dimensionless surface tension is set to $\Theta = 10$, and the dimensionless electrical work is set to $\Phi = 100$.

$S = 1$, one finds, accounting for the integer nature of cluster size, that $n_0 = 5$. In that case, the ionic pentamer, A_5^+ or A_5^- (equivalently, $I \cdot A_4$), is thermodynamically the most favored cluster size and would be predicted to exist, at equilibrium, in greater abundance than the bare ion.[3]

Thus, in the Thomson model, for the vast majority of substances, and except for extremely small saturation ratios, there exists what is known as a "stable prenucleus," an ionic cluster that consists of the ion with one or more attached neutral monomers.

Inspecting Figure 5.3, and analogous to the discussion in Sections 3.4 and 3.5, one can distinguish among three regimes, depending on the vapor saturation ratio.

For saturation ratios $S \leq 1$, the stable prenuclei have no tendency to grow, because as n increases above n_0, ΔG_n monotonically increases, without reaching a maximum. Thus, just as for the case of neutral self-nucleation, nucleation does not occur unless $S > 1$, in spite of the existence of stable prenuclei larger than the bare ion.

On the other hand, for the case $S > 1$, nucleation of larger particles does occur. Analogous to the situation of neutral self-nucleation discussed in Section 3.4.2, there exists some saturation ratio S^*, above which, as cluster size increases, ΔG_n monotonically decreases, without ever reaching a minimum. Thus, for $S > S^*$, there is no

---

[3] Note, however, that this does not imply that the ionic cluster of size n_0 is more abundant that the *neutral* monomer, because, depending on the degree of ionization, neutral monomers may be much more abundant than ionic monomers (i.e., bare ions).

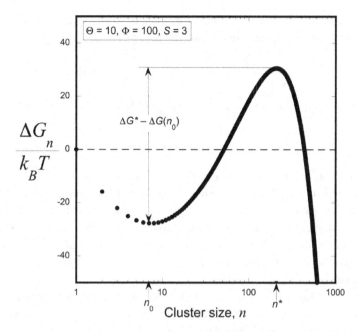

Figure 5.4 Illustration of energy barrier to nucleation in the classical Thomson model, for the case where the ion is of the same substance as the condensible vapor, and with $\Theta = 10$, $\Phi = 100$, and $S = 3$.

thermodynamic barrier to nucleation, and one is in the "collision-controlled" regime. For the conditions of Figure 5.3, that is evidently the case at $S = 10$. One notes in the curve for the $S = 10$ case the existence of an inflection point rather than a minimum in ΔG_n. Therefore, for $S > S^*$, stable prenuclei do not exist. Instead, cluster growth is irreversible, starting with the formation of the ionic dimer, and the nucleation rate is equal simply to the rate of the reaction between the bare ion and the neutral monomer, reaction (5.1).

For intermediate saturation ratios, $1 < S < S^*$, the ΔG_n curve exhibits both a minimum, at n_0, and a maximum, at the critical size n^*, as seen in Figure 5.3 for the cases $S = 3$ and $S = 4$. Thus one is here in the condensation/evaporation regime discussed in Chapters 1 and 2. In this regime, the energy barrier to nucleation is the difference $(\Delta G^* - \Delta G_{n_0})$, as illustrated in Figure 5.4 for the $S = 3$ case. Note that in the $S = 4$ case, even though ΔG_{n^*} is negative, an energy barrier to nucleation still exists, because $\Delta G^* > \Delta G_{n_0}$.

If one transforms equation (5.39) for ΔG_n from the discrete size variable n to an assumed continuous cluster radius r and defines a dimensionless cluster radius R by

$$R \equiv \frac{r}{r_1}, \tag{5.40}$$

then, as $n = R^3$ for assumed spherical clusters, equation (5.39) becomes

$$\frac{\Delta G_R}{k_B T} = \Theta\left(R^2 - 1\right) - \Phi\left(1 - R^{-1}\right) - \left(R^3 - 1\right)\ln S. \tag{5.41}$$

One can then find the location of both the stable prenucleus, R_0, and the critical size, R^*, by evaluating $d(\Delta G_R)/dR$ and setting the result to zero:

$$\frac{1}{k_B T}\frac{d(\Delta G_R)}{dR} = 2\Theta R - \Phi R^{-2} - 3(\ln S)R^2. \tag{5.42}$$

Setting the right-hand side to zero, multiplying both sides by R^2, and rearranging gives

$$(3\ln S)R^4 - 2\Theta R^3 + \Phi = 0. \tag{5.43}$$

This fourth-order equation is the dimensionless form of equation (5.30), for the case where the ion is the same size as the neutral monomer. The radii of the critical cluster and stable prenucleus are given by the largest and next largest roots of this equation. While the analytical solution is rather lengthy and cumbersome, it is worthwhile to recall that n is in fact an integer, and, as the value n_0 for which ΔG_n has its minimum is typically quite small, the values of n_0 and n^* are easily found by tabulating ΔG_n versus n and inspecting the result.

For the ionic cluster cases shown in Figure 5.3, one finds that the minimum in ΔG_n occurs for $S = 1$ at $n_0 = 5$; for $S = 3$ at $n_0 = 7$; and for $S = 4$ at $n_0 = 9$. For the $S = 3$ and $S = 4$ cases, n^* equals 208 and 94, respectively.

5.4   Steady-State Rate of Ion-Induced Nucleation in the Classical Model

Tohmfor and Volmer (Russell 1969; Tohmfor and Volmer 1938; Volmer 1939) used the Thomson model for ionic clusters to derive an expression for the steady-state rate of ion-induced nucleation. They assumed that the rate constants for monomer condensation onto ionic clusters were the same as onto neutral clusters. Treating cluster growth as commencing with stable prenuclei of size n_0, they followed the same mathematical derivation as in standard CNT to derive an expression for the steady-state nucleation rate.

Their resulting expression, converted to SI units, can be written as follows:

$$J_{\text{ion}} = \frac{\beta s^* N_{n_0}}{3n^*}\sqrt{\frac{4\pi(r^*)^2\sigma - \left(\dfrac{q^2}{4\pi\varepsilon_0 r^*}\right)(1-\varepsilon^{-1})}{\pi k_B T}}$$
$$\times \exp\left\{-\left[\frac{4\pi\sigma\left[(r^*)^2 - (r_{n_0})^2\right]}{3k_B T} - \frac{2q^2(1-\varepsilon^{-1})(1/r_{n_0} - 1/r^*)}{3k_B T(4\pi\varepsilon_0)}\right]\right\}, \tag{5.44}$$

where β is the same monomer flux as in equation (3.50) for neutral self-nucleation and N_{n_0} is the number density of the stable prenuclei. The theory typically assumes that N_{n_0} is equivalent to the ion number density N_i in the absence of clustering. In other words, it assumes that all bare ions grow to the size n_0 of the stable prenuclei before the clustering sequence begins. Indeed, if it is assumed that ions are preexisting, rather than continuously generated or otherwise introduced, then the population of stable prenuclei cannot exceed the initial population of ions.

Equation (5.44) can be simplified if one again assumes that the ions are of the same substance as the neutral monomers, so that n represents the number of monomers in a cluster, including the ion. (Again, this simplification also applies to the case where the ion, though of a different substance, has approximately the same radius as the neutral monomer, with the understanding that the index n now includes the ion.) In that case, transforming the cluster size variable from radius to number of monomers, and following the same relationships as in Section 5.3.3, one obtains for the steady-state ion-induced nucleation rate:

$$J_{\text{ion}} = \frac{\beta s_1 N_{n_0}}{3\sqrt{\pi}} \sqrt{\Theta - \frac{2\Phi}{n^*}} \exp\left\{ -\frac{\Theta}{3}\left[(n^*)^{2/3} - (n_0)^{2/3}\right] + \frac{4\Phi}{3}\left[(n_0)^{-1/3} - (n^*)^{-1/3}\right] \right\}.$$

(5.45)

It is interesting to compare the prefactor in front of the exponential with the corresponding term in equation (3.92) for the steady-state rate of neutral self-nucleation. The number density of stable ionic prenuclei is seen to play the same role in the prefactor as the number density of neutral monomers does in self-nucleation. This makes sense, as the clustering sequence in ion-induced nucleation is assumed to commence with monomer addition to the stable prenuclei. Otherwise, aside from the inclusion of the electrical work term in the square root, the prefactors are identical.

Equation (5.45) does not explicitly indicate a dependence on saturation ratio. That is because simple expressions are not possible for the values of n_0 and n^*, which *are* dependent on the saturation ratio, as is evident from Figure 5.3, for the cases where the saturation ratio lies in the condensation/evaporation regime. To use equation (5.45) to calculate the steady-state nucleation rate, one must first find the values of n_0 and n^*, for the given values of Θ, Φ, and S. Thus, the theory does not result in a simple closed-form expression for the steady-state rate of ion-induced nucleation, as it does, in equation (3.90), for single-component neutral CNT. However, as noted, it is straightforward to find n_0 and n^*, simply by tabulating $\Delta G_n(n)$.

5.5 Deficiencies of the Classical Model

The classical model of ion-induced nucleation has obvious deficiencies. It builds on the classical model of neutral self-nucleation and thus has all the deficiencies of that model, as discussed in Section 3.12. In addition, the Thomson model for the effect of the ion on the free energy of cluster formation is obviously a crude approximation when applied to small clusters, especially considering the added chemical complexity when the ion is of a different chemical substance than the condensible vapor. Furthermore, the model for condensation rate coefficients is even cruder than in the CNT of neutral self-nucleation, because the effect of charge on collision rates between ionic clusters and neutral monomers is neglected. However, it is known that ion-neutral collisions are affected by Langevin dynamics, where the ion induces an electric dipole in the neutral that results in an attractive force that modifies collision trajectories and increases collision rates. As discussed in Section 9.3.3, the resulting collision rate coefficient depends on the polarizability of the neutral monomer, which is substance-dependent. For all these reasons, if CNT of neutral self-nucleation has been found, in general, to be in poor agreement with experiments, one can expect, in general, that the classical model of ion-induced nucleation would be in even worse agreement.

One qualitative question that has been examined experimentally is whether the nucleation rate is affected by the ion polarity. In the Thomson model for the free energy of formation of ionic clusters, equation (5.25), as well as in the result for the steady-state nucleation rate, equation (5.44), the cluster charge appears only in the term q^2. Therefore, according to the model, the nucleation rate should be the same regardless of whether ions are positive or negative, all else being equal.

However, most experiments, reviewed in Keshavarz et al. (2020), have found that there is indeed a sign preference, with some studies concluding that positive ions are more effective in promoting nucleation, while other studies, beginning with Wilson's cloud chamber experiments (Wilson 1897), have found that negative ions are more effective. A limitation of most of these studies is that it is experimentally difficult to control the size and chemical composition of the ions and to have the ion size and chemical composition be identical in experiments that compare the nucleation of a given vapor induced by either unipolar positive or unipolar negative ions.

Keshavarz et al. (2020) performed quantum chemical calculations to explain the positive ion preference for nucleation of diethylene glycol induced by several different ions of controlled size and composition. On the other hand, theoretical studies of ion-induced nucleation in other systems, for example, pure water clusters and binary sulfuric acid–water clusters (Nadykto et al. 2008), have found a negative ion preference, the latter explained in terms of acid–base interactions.

The preponderance of work addressing this question indicates that ion-induced nucleation in most chemical systems is indeed affected by ion polarity, contrary to the Thomson model. Whether anions or cations are more effective at inducing nucleation appears to depend on the specific chemical system involved.

Homework Problems

5.1   Consider ionic H_2O clusters in the Thomson model, with the ion being $(H_2O)^+$ or $(H_2O)^-$. For a temperature of 20°C and for saturation ratios of 1, 3, and 10, prepare a table of $\Delta G_n(p_1)/k_B T$ and graph the results, as in Figure 5.3.

Based on the table, what are the sizes of the stable prenucleus, n_0, and the critical cluster, n^*, for the case $S = 3$?

What is the value of the nucleation energy barrier for saturation ratios of 3 and 10?

5.2   Using the table you constructed in problem 5.1, again consider water at 20°C. Suppose that 0.1% of the vapor phase water molecules are ionized and that 100% of these ions grow to form stable prenuclei of size n_0. For saturation ratios of 3 and 10, calculate and compare the steady-state nucleation rates for both ion-induced nucleation and neutral self-nucleation, as predicted by classical theory.

5.3 Again for ionic water clusters at 20°C, find the value of S^* (to three significant figures), that is, the value of saturation ratio above which there is no thermodynamic barrier to ion-induced nucleation.

5.4 Solve equation (5.30) numerically to find the expressions for the roots that give the extrema of ΔG_n.

6 Atomistic Approaches to Homogeneous Nucleation

6.1 Similarities and Differences between Atomistic and Classical Theories

The classical model for single-component homogeneous nucleation presented in Chapter 3 results in a closed-form analytical expression for the steady-state nucleation rate. Its shortcomings arise primarily from the crudeness of the liquid droplet model to describe small clusters, leading to errors in the nucleation rate that may be as large as many orders of magnitude, as well as incorrectly predicting the dependence of nucleation rate on both temperature and saturation ratio.

In this chapter, we consider the use of atomistic data to potentially improve this situation. By "atomistic data," we mean values for free energies and rate constants, in which a cluster is treated as a distinct molecular species, rather than as a small piece of the bulk condensed phase. While appropriate atomistic data on cluster properties for most substances remain sparse, one can anticipate that they will become increasingly available. They may in general be determined experimentally, theoretically, or by means of computational chemistry.

Classical theory divides the free energy of a cluster into "volume" and "surface" contributions. In an atomistic theory, this distinction is not meaningful, especially for the small clusters, usually involving fewer than 100 atoms or molecules, that are of interest in nucleation. With reference to Figure 1.4, for such small clusters, a large fraction of atoms lie on the "surface." Given atomistic data, it may be of interest to infer equivalent values of properties such as surface tension, but for small clusters such an exercise essentially involves the construction of an artifice to facilitate comparisons with classical theory.

In this chapter, we continue to assume that single-component homogeneous nucleation follows the scenario outlined in Chapter 2, as given by the reaction sequence (2.1)–(2.4). In this regard, two important distinctions between the classical and atomistic approaches should be noted. First, classical nucleation theory (CNT) assumes that clusters are structureless spheres, as would usually be a reasonable approximation for liquid droplets. However, in reality clusters may have structure and may be far from spherical, and in many cases, different structures, or isomers, can exist for clusters having identical numbers of atoms and the same elemental composition. For example, Figure 3.1 shows the four most stable isomers predicted for $(H_2O)_{10}$ (Maheshwary et al. 2001), and Figure 6.1 shows a number of different isomers that have been predicted to exist for the water heptamer, $(H_2O)_7$ (Shields et al. 2010). An approach often taken is that different isomers, as with different rotational, vibrational, and electronic energy

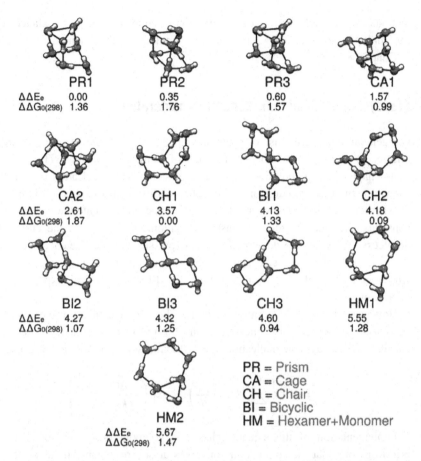

	PR1	PR2	PR3	CA1
$\Delta\Delta E_e$	0.00	0.35	0.60	1.57
$\Delta\Delta G_{0(298)}$	1.36	1.76	1.57	0.99

	CA2	CH1	BI1	CH2
$\Delta\Delta E_e$	2.61	3.57	4.13	4.18
$\Delta\Delta G_{0(298)}$	1.87	0.00	1.33	0.09

	BI2	BI3	CH3	HM1
$\Delta\Delta E_e$	4.27	4.32	4.60	5.55
$\Delta\Delta G_{0(298)}$	1.07	1.25	0.94	1.28

	HM2
$\Delta\Delta E_e$	5.67
$\Delta\Delta G_{0(298)}$	1.47

PR = Prism
CA = Cage
CH = Chair
BI = Bicyclic
HM = Hexamer+Monomer

Figure 6.1 Lowest energy configurations of $(H_2O)_7$, predicted by high-level quantum chemistry calculations of Shields et al. (2010). Reprinted with permission from Shields et al. (2010). Copyright 2010 American Chemical Society.

states, are treated using the tools of statistical mechanics to produce appropriately ensemble-averaged thermodynamic properties. This approach is appropriate if the relative abundances of the different isomers are equilibrated among themselves. If that assumption is not correct, or at least if departures from this equilibrium are significant, then each isomer must be treated as a separate chemical species in a kinetic scheme, which would include the rates of interconversion among at least the most important isomers, as discussed in Chapter 8 on chemical nucleation.

Second, the use of atomistic data may call into question the assumption that clusters grow by monomer addition only, as in the reaction sequence (2.1)–(2.4). Clusters of many substances have been found to exhibit magic number effects, meaning that clusters of certain sizes (the magic numbers) are much more abundant, at equilibrium, than those of other sizes. If conditions exist where the abundance of magic-numbered clusters becomes comparable to that of the monomer, then the contribution to the growth of cluster–cluster reactions of the form given by equation (2.5) is likely to become important. In such cases, the theory presented in Chapter 2 would not apply, and nucleation would best be treated by a set of coupled population balance equations for multiple species and reactions, not limited to the monomer, as discussed in Chapter 8.

6.2   Steady-State Nucleation: General Considerations

On general considerations, one can anticipate a few basic features regarding the differences and similarities between the use of atomistic data and CNT.

In Chapter 2, an expression for the rate of steady-state nucleation is derived in the form of a summation, equation (2.68). The validity of this expression is independent of whether one uses CNT or an atomistic approach to evaluate the condensation rate constants and free energies of cluster formation. In Chapter 3, the classical liquid droplet model is used to develop a simple expression for the free energy of formation as a function of cluster size and thereby to convert the summation to an integral, which is then evaluated to produce a closed-form analytical expression for the nucleation rate.

However, if one has atomistic data for the size-dependent values of ΔG_n^0 and condensation rate constants, then equation (2.68) provides a straightforward means to calculate the steady-state nucleation rate, without resorting to the liquid droplet model:

$$ J = N_1{}^2 \left\{ \sum_{n=1}^{M} \left[k_{n+1} \left(\frac{p_1}{p^0} \right)^{n-1} \exp\left(-\frac{\Delta G_n^0}{k_B T} \right) \right]^{-1} \right\}^{-1} . \qquad (2.68) $$

Implementation of this equation has two potential problems. First, it is not a priori obvious that the summation converges at any reasonably finite value of n, for arbitrary values of ΔG_n^0 and the forward rate constants. However, as shown in Section 6.3, the summation can be expected in most cases of interest to converge quite satisfactorily.

Second, even if atomistic data exist, they can be expected to exist only up to some finite size, which may be smaller than the critical size under given conditions. To address this issue, one might employ a hybrid of atomistic and classical approaches, together with experimental data if these were available. If measurements of homogeneous nucleation rates are available for a range of saturation ratios along an isotherm, then these data could be used to estimate both the critical size, from the nucleation theorem, equation (3.103), and the value of ΔG^* evaluated at the temperature and saturation ratio of the experiments. Measurements made for different ranges of saturation ratio could then be used to fill in the table of ΔG_n^0 for the corresponding values of n, interpolating as necessary, and an $n^{2/3}$ scaling could be used, as in equation (3.29), to extrapolate to larger sizes.

Regarding condensation rate constants, as discussed in Section 3.6.1, CNT treats collisions of monomers with clusters as if the clusters were very large, neglects the effects of interaction potentials that might alter collision rates, and assumes unity sticking coefficients. In any atomistic theory, the first of these assumptions is easily dispensed with. Instead of using equation (3.54), as in CNT, one can use equation

(3.44), the bimolecular collision rate from hard-sphere gas kinetic theory, which accounts for the motion of the cluster. The difference is not negligible, as shown in Figure 3.5.

Regarding the other two assumptions, the existence of attractive interaction potentials between monomers and clusters could increase collision rates, while chemical reaction energy barriers could cause sticking coefficients to be less than unity. Both of these assumptions could be examined, for specific substances, by computational chemistry, and the resulting values of k_n could be used directly in equation (2.68).

6.3 Convergence of the Summation Expression

Equation (2.68) is a useful form of the summation expression, as atomistic data for pressure-dependent thermodynamic properties such as entropy and Gibbs free energy are generally given at standard reference pressure p^0, conventionally 1 atm or 1 bar. Furthermore, equation (2.68) directly uses the actual monomer partial pressure p_1 and requires no knowledge of the equilibrium vapor pressure of the substance. This expression, then, provides a direct recipe for inserting such data.

Writing the summation in equation (2.68) term by term, one has

$$\frac{J}{N_1^2} = \left[\frac{1}{k_2} + \frac{\exp\left(\frac{-\Delta G_2^0}{k_B T}\right)}{k_3(p_1/p^0)} + \frac{\exp\left(\frac{-\Delta G_3^0}{k_B T}\right)}{k_4(p_1/p^0)^2} + \cdots + \frac{\exp\left(\frac{-\Delta G_M^0}{k_B T}\right)}{k_{M+1}(p_1/p^0)^{M-1}}\right]^{-1}. \tag{6.1}$$

6.3.1 Convergence of Summation in CNT

It is impossible to know whether this expression converges without knowing the size-dependent behavior of the cluster free energies and condensation rate constants. In Section 3.12.2, we show that if one assumes CNT values for the rate constants and free energies of formation, the summation expression is overwhelmingly likely to converge within a quite tractable number of terms – only slightly greater than the critical size – at least for cases where nucleation rates are not negligible. That in itself does not prove that the summation converges if one uses atomistic data for rate constants and free energies. However, even though CNT may be seriously in error at very small cluster sizes, we expect it at least to provide qualitative guidance for the behavior of the terms in the summation as cluster size becomes large and the validity of the liquid droplet model improves.

6.3.2 Convergence with Atomistic Data

Does the conclusion of the previous paragraph imply that the summation would converge if atomistic data were used instead of CNT? The answer is yes. The reason is that, within the condensation/evaporation regime, that is, for values of saturation ratio $1 < S < S^*$, there must exist at least one critical size that constitutes the barrier(s) to nucleation.

As given by equation (2.80), the nucleation rate can be written in terms of the free energy of cluster formation evaluated at the partial pressure p_1 of the monomer vapor. In the classical model, for values of saturation ratio within the condensation/evaporation regime, the curve of free energy versus cluster size (Figure 3.3) is unimodal, that is, it has a single maximum. As suggested by Figure 3.24, it is a priori possible for atomistic data to exhibit multiple local maxima in the curve of ΔG_n versus n, but finally there must exist a size beyond which further growth is "downhill" in free energy space, because for saturation ratios greater than unity the bulk condensed phase, coexisting with saturated vapor, constitutes the stable equilibrium of the system. Mathematically, the existence of a maximum in $\Delta G_n(p_1)$ is guaranteed, within the condensation/evaporation regime, by the $(n-1)k_B T \ln S$ term in equation (2.79), which is valid regardless of whether ΔG_n is obtained from CNT or atomistically. Therefore, notwithstanding possible size dependencies of the condensation rate constant, one can again expect the summation to converge to within reasonable numerical criteria within some distance not far past the critical size.

At sufficiently large cluster sizes, the qualitative trends described by CNT should hold, that is, the stepwise free energy change given by equation (3.23) should be valid, as long as surface tension is appropriately evaluated.

However, even if the stepwise free energy change in this "large cluster" regime is identical to that given by CNT, this does not imply that atomistic data will yield the same result as CNT for the nucleation rate.

There are two reasons for this. First, in many cases the critical size will lie well below this large cluster regime. Second, by equation (2.26), ΔG_n equals the sum of all the stepwise free energy changes for the sequence of condensation events that grow a cluster, starting from the monomer. Therefore, differences between atomistic data and CNT in the values of $\Delta G_{n-1,n}$, which are expected at small cluster sizes, will affect the value of ΔG_n at larger sizes, even at sizes for which the stepwise free energy changes are well described by CNT.

For very small clusters, it is often meaningless to divide the cluster into "interior" and "surface," but as the cluster size increases the distinction becomes meaningful, and the structure and properties of the interior of the cluster must finally approach those of the bulk condensed phase, whether liquid or solid. One can then partition the cluster free energy into contributions from the volume and surface free energies, as in equation (3.2), and, following the same thermodynamic arguments as in Section 3.3.3, but without invoking the liquid droplet model, write, for the stepwise free energy change,

$$\Delta G_{n-1,n}(p_1) = \tilde{\mu}_c(p_s) - \tilde{\mu}_v(p_s) + \sigma(s_n - s_{n-1}) - k_B T \ln S, \qquad (6.2)$$

where the subscript "c" denotes the bulk condensed phase (not necessarily liquid) to which the cluster's structure corresponds, p_s is the equilibrium vapor pressure above that bulk condensed phase, and σ is the cluster's surface tension (i.e., surface energy per unit area, not necessarily the same as in the liquid droplet model). In that case, thermodynamics requires $\tilde{\mu}_c(p_s) = \tilde{\mu}_v(p_s)$, and the expression for the stepwise free energy change reduces to equation (3.22).

6.4   Examples of Atomistic Data on Free Energies of Cluster Formation

Two examples are presented from the literature on atomistic data for free energies of cluster formation in single-substance homogeneous nucleation: water clusters, up to size $(H_2O)_{10}$, and aluminum clusters, up to size Al_{60}. Both sets of data were obtained by means of high-level quantum chemistry computational methods. As the detailed theory of such calculations is beyond the scope of this text, the interested reader is referred to the references cited in Sections 6.4.1 and 6.4.2 for more information.

6.4.1   Water Clusters

Temelso et al. (2011) calculated standard stepwise free energy changes for water clusters by using an approach that combined molecular dynamics sampling with high-level ab initio calculations. For each cluster, they considered global and many low-lying local energy minima configurations and accounted for anharmonicity of vibrational energy modes. They then used their results for the stepwise free energy changes, $\Delta G^0_{n-1,n}$, in equation (2.26) to calculate the cumulative free energy of formation, ΔG^0_n, for n up to 10, assuming a standard pressure of $p^0 = 1$ atm. They tabulated their results for 10 temperatures in the range of 0–373.15 K.

Figure 6.2 shows the calculated values of Temelso et al. of the dimensionless stepwise free energy change at 250 K, adjusted to a saturation ratio of unity by means of equation (2.82), compared to the values given by CNT. The latter are based on a value for the dimensionless surface tension Θ, for water at 250 K, of 10.7. In making this comparison, we take the equilibrium vapor pressure of water at 250 K to equal 95.1 Pa. Thus $p^0 = 1$ atm corresponds to a saturation ratio of 1,051. From the discussion in Section 3.3.3, the case $S = 1$ is the most logical condition for the comparison of atomistic data and CNT, as for any other saturation ratio the results for $\Delta G_{n-1,n}/k_BT$ for both CNT and atomistic data are obtained simply by subtracting $(n-1)\ln S$.

As is evident from Figure 6.2, the atomistic values for the stepwise free energy changes differ substantially from those predicted by CNT. Most of the atomistic values are higher than in CNT, while, on the other hand, two of the atomistic values for stepwise free energy change are lower than in classical theory. Moreover, the atomistic values show a trend for stepwise free energy change versus cluster size that is highly size dependent and non-monotonic, unlike the smooth monotonic trend predicted by CNT.

Figure 6.3 shows a comparison of the free energies of cluster formation $\Delta G_n/k_BT$, evaluated at $S = 1$, using either the atomistic data of Temelso et al. (2011) or self-consistent classical theory.[1] This figure makes evident the cumulative effects of the generally larger stepwise free energy changes in the atomistic data compared to classical theory. Already by size $n = 10$ the value of ΔG_n calculated by Temelso et al. substantially exceeds (by 12.3 k_BT, or 6.10 kcal · mol^{-1}) the value predicted by CNT. If this

---

[1] Note that in making this comparison it would make little sense to use the standard version of classical theory, which erroneously sets the free energy of monomer formation to a nonzero value.

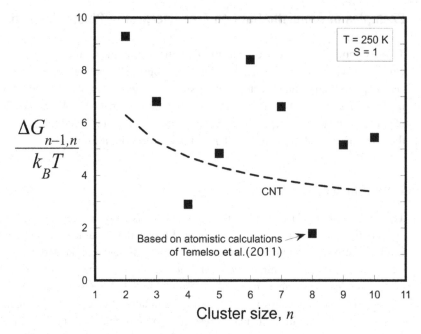

Figure 6.2 Stepwise free energy change of water clusters at 250 K and $S = 1$, based on the atomistic calculations of Temelso et al. (2011), compared to values from classical theory. Adapted from Girshick (2014) with the permission of AIP Publishing.

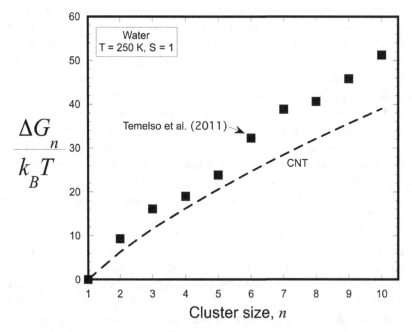

Figure 6.3 Free energies of formation of water clusters at 250 K and $S = 1$, based on the atomistic calculations of Temelso et al. (2011), compared to values from self-consistent classical theory. Adapted from Girshick (2014) with the permission of AIP Publishing.

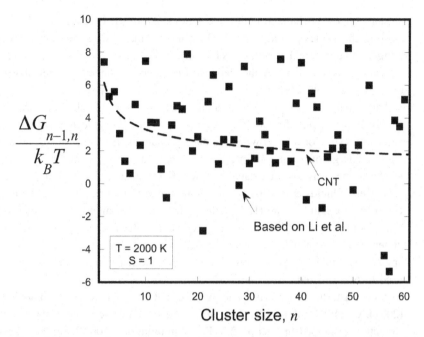

Figure 6.4 Stepwise free energy change of aluminum clusters at 2,000 K and $S = 1$, based on the atomistic calculations of Li et al. (2007), compared to the values from classical theory. Adapted from Girshick (2014) with the permission of AIP Publishing.

result is at least qualitatively correct, then one expects the actual value of ΔG_n to be larger than that predicted by CNT even for clusters much larger than size 10, and therefore the steady-state nucleation rates for water at 250 K should be much lower than predicted by CNT, even if the critical size is much larger than 10. Indeed, experimental results in the literature show that to be the case (Brus et al. 2009).

6.4.2   Aluminum Clusters

Li et al. (2007) calculated the stepwise free energy changes for monomer condensation onto aluminum clusters, $\Delta G^0_{n-1,n}$, for all sizes up to Al_{60}, assuming a standard pressure of 1 atm. Their Table 10 tabulates these values in $kcal \cdot mol^{-1}$, for temperatures of 1,500, 2,000, 2,500, and 3,000 K. Figure 6.4 shows their data for the stepwise free energy change of aluminum at 2,000 K, adjusted to a saturation ratio of unity using equation (2.82).[2] Also shown are the values predicted by CNT.[3]

As with water, the stepwise free energy changes for aluminum calculated atomistically are seen to be highly non-monotonic, and this behavior persists up to the largest

---

[2] The equilibrium vapor pressure of aluminum at 2,000 K equals approximately 648 Pa (Alcock et al. 1984). Thus 1 atm corresponds to a saturation ratio of approximately 156.

[3] The CNT calculations for aluminum assume a dimensionless surface tension at 2,000 K of $\Theta = 10.46$, based on the measurements of Sarou-Kanian et al. (2003).

cluster size, 60, considered in the calculations. Atomistic data for a number of sizes lie close to the values predicted by CNT, while for many sizes the deviations are large, lying both above and below the CNT values. Indeed, for several sizes the atomistic data indicate negative values of the stepwise free energy change, a feature that is qualitatively at odds with CNT at $S = 1$.

Additionally, Li et al. (2007) tabulate (their Table 11) the standard Gibbs free energy of formation for all sizes up to Al_{60}, *including the monomer*, at the same temperatures as for the stepwise free energy changes. Here it is important to note that the free energy of formation for Al clusters tabulated by Li et al. (2007) is not defined in the same way as that tabulated for H_2O clusters by Shields, Temelso, and coworkers. Shields et al. (2010) refer to the property they tabulate as "the total change G^0 for the formation of an $(H_2O)_n$ cluster from n water monomers," which is identical to the definition of ΔG_n^0 used in this text. Hence, the free energy of formation of the dimer, ΔG_2^0, is equal to the stepwise free energy change $\Delta G_{1,2}^0$, and equation (2.26) applies. The free energy of formation of the monomer is implicitly zero and does not need to be tabulated.

In contrast, the standard free energy of formation of Al clusters tabulated by Li et al. (2007), which they denote by $\Delta_f G^0(n)$, is the standard thermochemical free energy of formation discussed in Section 2.3. For aluminum at 1,500, 2,000, and 2,500 K, the "standard reference element" is liquid-phase monatomic aluminum. Thus $\Delta_f G_1^0$ at these temperatures does not equal zero but rather equals the free energy change per atom for the conversion of aluminum from liquid to vapor, at the temperature in question, and this free energy change must be counted for every monomer composing the cluster.[4] Thus, the values of $\Delta_f G^0(n)$ tabulated by Li et al. (2007) can be calculated from their stepwise free energy changes using[5]

$$\Delta_f G_n^0 = \sum_{i=2}^{n} \Delta G_{i-1,i}^0 + n \Delta_f G_1^0, \quad n \geq 2. \tag{6.3}$$

Care must thus be taken regarding which definition of the free energy of cluster formation is used in a particular data set. For nucleation calculations, the safest approach is to use equation (2.26), together with values of the stepwise free energy changes if these are available. On the other hand, if the stepwise free energy changes are not available, but rather data on $\Delta_f G_n^0$ that are based on the standard thermochemical definition of free energy of formation, then equations (6.3) and (2.26) can be used to yield

---

[4] At temperatures above 2,790.812 K, the vapor phase becomes the stable form of aluminum at 1 bar (Chase 1998), and thus $\Delta_f G_1^0$ does equal zero at these temperatures.

[5] Actually Li et al. (2007) use an experimental measurement for the free energy of formation of the Al dimer (as well as for the monomer). This measurement does not agree exactly with equation (6.3) for the dimer, based on their numerical calculation of $\Delta G_{1,2}^0$. Thus instead of using equation (6.3) they use the recursion relation,

$$\Delta_f G_n^0 = \Delta G_{n-1,n}^0 + \Delta_f G_{n-1}^0 + \Delta_f G_1^0,$$

where they use experimental values for both the monomer and dimer, and their numerical calculations for all larger sizes.

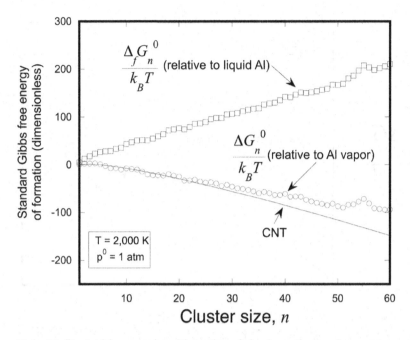

Figure 6.5 Standard free energies of formation of aluminum clusters. Squares represent the values from Li et al. (2007), which are standard free energies of formation defined in the usual thermochemical manner, relative to the stable form of Al at 2,000 K, which is liquid. Circles represent the standard free energy of formation relative to Al vapor, based on equation (2.26) and using the Li et al. (2007) data for stepwise free energy change. The curve for classical theory uses equation (3.27), with S set to 156, corresponding to $p_1 = 1$ atm.

$$\Delta G_n^0 = \Delta_f G_n^0 - n\Delta_f G_1^0. \tag{6.4}$$

Figure 6.5 illustrates this conversion. It shows the Li et al. (2007) data for $\Delta_f G_n^0$ at 2,000 K, the corresponding values of ΔG_n^0 obtained using equation (6.3), and the values of ΔG_n^0 based on self-consistent CNT. Note that negative values of ΔG_n^0, for both the atomistic values and CNT, are a consequence of the high saturation ratio, about 156, of Al vapor at 2,000 K and 1 atm.

With ΔG_n^0 calculated from atomistic data, the values of ΔG_n at $S = 1$ can then be determined from equation (2.78), where in this case $p_1 = p_s$. The results are shown in Figure 6.6, again compared to self-consistent CNT. An initially surprising aspect of this comparison is that the values of ΔG_n predicted by CNT at $S = 1$ appear to lie remarkably close to the atomistic values up to $n \approx 25$ but then begin increasingly to deviate at larger sizes. This impression of good agreement at small sizes is misleading. From Figure 6.4, the discrepancies between the atomistic data and CNT for the *stepwise* free energy changes at small sizes are large, but these discrepancies tend in this case to offset each other in taking the sum, up to around size 25, via equation (2.26), to obtain the free energy of cluster formation.

Aside from that, at $S = 1$ the critical size is infinite – that is, no maximum exists in ΔG_n – and thus the ordinate scale in Figure 6.6 tends to obscure the important discrepancies in ΔG_n that would be evident at higher saturation ratios. This is illustrated in

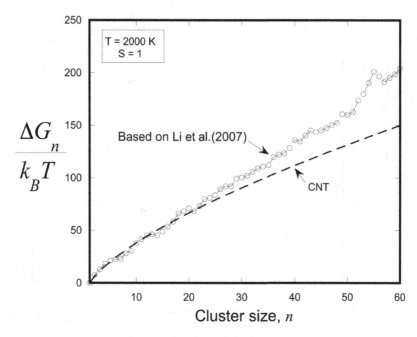

Figure 6.6 Free energies of formation of aluminum clusters at 2,000 K and $S = 1$, based on atomistic calculations of Li et al. 2007, compared to the values from self-consistent classical theory. Adapted from Girshick et al. (2009) with the permission of AIP Publishing.

Figure 6.7, which shows ΔG_n obtained from either the Li et al. (2007) data or CNT at 2,000 K and a saturation ratio of 20, which is a more interesting saturation ratio for nucleation, as it lies in the condensation/evaporation regime.

Here it is seen that the atomistic data and CNT have profound qualitative differences. The atomistic data for ΔG_n exhibit a highly non-monotonic behavior, with numerous local maxima and minima, and the sharply defined critical size seen in the atomistic data at $n = 55$ (assuming that no higher maximum exists beyond size 60, which is not known)[6] is remarkably different than the critical size predicted by CNT, which from equation (3.32) equals approximately 12.6. Furthermore, the energy barrier to nucleation predicted by the atomistic data is much larger than in CNT.

The minima in ΔG_n seen in Figure 6.7 can be said to correspond to metastable equilibrium states. They are stable in the sense that at these minima both growth and decay of an n-mer would cause ΔG_n to increase and thus are thermodynamically inhibited, but they are metastable as they are local not global minima.

---

[6] For some chemical substances, for example, carbon fullerenes or gold clusters, magic numbers are known to exist up to much larger sizes, implying the existence at large sizes of local minima and corresponding local maxima in ΔG_n.

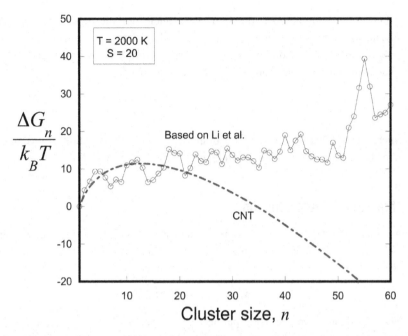

Figure 6.7 Free energies of formation of aluminum clusters at 2,000 K and $S = 20$, based on the atomistic calculations of Li et al. (2007), compared to the values from self-consistent classical theory. Adapted from Girshick et al. (2009) with the permission of AIP Publishing.

6.5   Condensation Rate Constant

The condensation rate constant for real systems undergoing single-component homogeneous nucleation may differ from the value assumed by CNT. From equation (3.54), the rate constant in CNT for condensation of a monomer on an n-mer can be written as

$$k_{n+1} = \sqrt{\frac{k_B T}{2\pi m_1}} s_1 n^{2/3}. \tag{6.5}$$

There are several ways in which an atomistic model for nucleation could improve on this expression.

6.5.1   Bimolecular Collision Rate from Hard-Sphere Kinetic Theory

Equation (6.5) is based on the monomer flux β to an infinite plane, equivalent to assuming that the n-mer is so much larger than the monomer that the n-mer is effectively a stationary target. As seen in Figure 3.5, this approximation underpredicts the monomer–cluster collision frequency by a factor that ranges from ~4 for monomer–dimer collisions[7] to ~33% for collisions between a monomer and a 100-mer.

---

[7] For monomer–monomer collisions, CNT underpredicts the collision rate by a factor of ~2.8 not ~5.6 as suggested by Figure 3.5, because the factor of 2 in equation (3.55) must be applied.

In using the summation expression for the nucleation rate, equation (2.68), there is no advantage in using this approximation. At the least, even if one still assumes spherical, nonattractive clusters, one can use the more accurate expression given by the bimolecular collision rate from hard-sphere gas kinetic theory,

$$k_{n+1} = \frac{\beta_{1,n}}{\varsigma},$$
(6.6)

where the collision frequency function $\beta_{1,n}$ is given by equation (3.45), and ς equals 2 for the case $n = 1$ (monomer–monomer collisions) and unity otherwise.

6.5.2   Three-Body Recombination

As noted in Section 2.1, monomer condensation is not an elementary reaction but requires a third body to stabilize the larger cluster formed. Thus the condensation/ evaporation reaction should properly be written as

$$A_{n-1} + A_1 + M \leftrightarrows A_n + M,$$
(6.7)

where M denotes any collision partner.

Here the forward reaction represents the sequence of (at least) two elementary reactions,

$$A_{n-1} + A_1 \rightarrow A_n^*$$
(6.8)

and

$$A_n^* + M \rightarrow A_n + M,$$
(6.9)

where A_n^* represents an excited complex, or transition state, that carries excess energy from the collision. As a cluster grows, its number of internal degrees of freedom rapidly increases, and the excess energy can be absorbed by internal rearrangement of the cluster's structure. For very small clusters, especially those containing approximately five or fewer monomers, which have fewer degrees of freedom, this may not be possible. In that case, the excited complex A_n^* may spontaneously dissociate, within one vibrational period, unless a third body M is available to stabilize the excited complex by absorbing its excess energy in a collision.

Depending on the species involved, calculations or measurements for the forward rate constants of reactions (6.7) and (6.8) may exist in the literature. Let these rate constants be denoted by $k_{n(M)}$ and k_n^*, respectively.

The rate constant $k_{n(M)}$, which has units of $m^6 \cdot s^{-1}$, is defined with respect to the overall three-body reaction (6.7):

$$R_n = k_{n(M)} N_{n-1} N_1 N_M.$$
(6.10)

This rate constant is not itself usable in the summation expression for the nucleation rate, whose derivation assumes that all reactions in the clustering sequence are of the form of reaction (2.4), with the forward rate constant defined in terms of equation (2.12).

In general, the effectiveness of a third body in stabilizing A_n^* depends on the chemical species involved, in which case one must write reaction (6.7) separately for each component in a multicomponent gas as the third body, where the values of $k_{n(M)}$ may differ for each of these reactions. For simplicity, however, let us assume that the nucleating system consists of either a single component (the condensing vapor) or a dilute vapor in an inert carrier gas, or that all third bodies are equally effective.

In all these cases, recognizing that one can write

$$N_M = \frac{p}{k_B T}, \tag{6.11}$$

where p is the total pressure; one can define an "effective two-body rate constant" by

$$k_n = \left(\frac{p}{k_B T}\right) k_{n(M)}. \tag{6.12}$$

Studies that report the rate constant for reaction (6.8) may also report characteristic lifetimes, τ_{lifetime}, for the excited complex A_n^* before its spontaneous dissociation. In that case, assuming that this lifetime is shorter than the characteristic time $\tau_{\text{collision}}$ between collisions with any stabilizing third body, one can write

$$k_n \approx k_n^* \left(\frac{\tau_{\text{lifetime}}}{\tau_{\text{collision}}}\right). \tag{6.13}$$

As one example, Martin et al. (1990) reported ab initio calculations of the rate constant $k_{2(M)}$ of three-body recombination, reaction (6.7), to form Si_2 at temperatures of 800, 1,000, and 1,200 K,

$$Si + Si + M \rightarrow Si_2 + M, \tag{6.14}$$

with either Si or Ar as the third body, and Gai et al. (1988) reported calculations of k_3^* for formation of the silicon trimer via

$$Si_2 + Si \rightarrow Si_3^*, \tag{6.15}$$

at temperatures in the range 100–1,500 K, together with a range of average lifetimes, from $\sim 5 \times 10^{-13}$ to 5×10^{-11} s, for the excited complex in the simulations.

These data were used by Kelkar et al. (1996) in a study of nucleation from supersaturated silicon vapor. Their resulting estimates for the effective rate constants k_n of reaction (2.4) for silicon are shown in Figure 6.8, with total pressure assumed to equal 1 bar, temperature 1,000 K, and with argon as the third body. Here k_2 was obtained using the value of $k_{2(M)}$ from Martin et al. (1990), together with equation (6.12), and k_3 was obtained using the value of k_3^* from Gai et al. (1988), together with equation (6.13), assuming a value of 5 ps for the lifetime of the excited trimer. For condensation to form clusters of sizes $n \geq 6$, it was assumed that for clusters this large the number of internal degrees of freedom is large enough to accommodate the excess energy of collision and that the rate constant is given by the hard-sphere collision frequency function, equation (3.45). Finally, the values shown for the rate constants to form the tetramer and pentamer were obtained by multiplying their associated collision frequency functions by arbitrary sticking coefficients (see equation (3.36)) of 0.5 and 0.9, respectively, so as to produce the smooth interpolation seen in Figure 6.8.

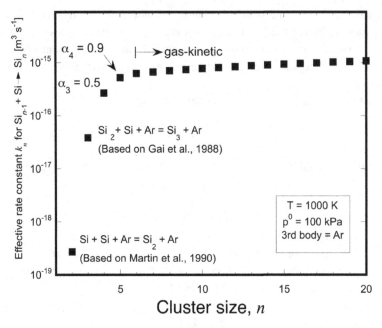

Figure 6.8 Estimated effective two-body rate constants k_n for the formation of Si clusters. "α" denotes the sticking coefficient. Adapted from Kelkar et al. (1996) with the permission of Elsevier.

If the total pressure is sufficiently high that the characteristic time for molecular collisions is shorter than the characteristic lifetime of excited transition complexes, particularly of the smallest clusters, then one is in the "high-pressure limit." In this case, the third body M can be omitted from all reactions of the form of equation (6.7), and they can be treated for purposes of nucleation calculations as if they were elementary reactions. At atmospheric pressure, this is often, though not always, the case.[8] Since the rate of collisions of all molecules with a single cluster is proportional to the total molecular number density, the need to account for the third body in equation (6.7) becomes more important as pressure is reduced.

Reduction in effective two-body rate constants by third-body effects for small clusters may or may not affect the steady-state nucleation rate. If the reduction is strong, as for dimers in Figure 6.8, then the $n = 2$ term in the summand of equation (2.68) may be larger than the term for the critical cluster size. In that case, the kinetic bottleneck for dimerization would be stronger than the free energy barrier for forming critical clusters, and dimerization would be rate limiting. Moreover, even if this were not the case – that is, even if the steady-state nucleation rate is not altered – the time required to reach steady state can be strongly affected by reduced rates of cluster formation at sizes much smaller than the critical size. In such cases, depending on the conditions, steady-state nucleation may not be a valid assumption. This point is discussed further in Section 7.3.4.

[8] As seen in Figure 6.8, it is evidently not the case for silicon nucleation from dilute silicon vapor in argon at 1 bar and 1,000 K.

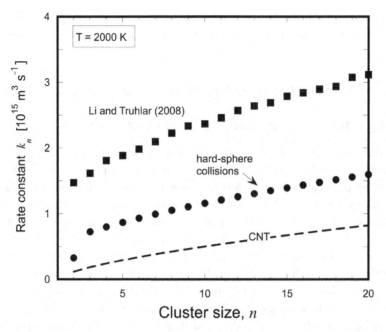

Figure 6.9 Rate constants for $Al_{n-1} + Al \rightarrow Al_n$, from Li and Truhlar (2008), hard-sphere kinetic theory, and classical nucleation theory.

6.5.3   Condensation Rate Constants with Attractive Intermolecular Forces

The collision frequency function given by equation (3.45) neglects short-range attractive intermolecular forces between collision partners, which can alter their trajectories so as to increase collision rates and hence condensation rate constants. Li and Truhlar (2008) used molecular dynamics trajectory simulations to calculate k_n for aluminum clusters up to $n = 60$. Their results up to $n = 20$ are shown in Figure 6.9, together with values from both CNT and hard-sphere kinetic theory.

As discussed in Section 3.7.1, hard-sphere gas kinetic theory gives higher rate constants than CNT, because CNT neglects the finite size and motion of the cluster onto which the monomer condenses. As seen in Figure 6.9, the Li and Truhlar (2008) calculations predict rate constants that are yet higher, by a factor of about two compared to hard-sphere kinetic theory.[9]

Thus, even though errors in rate constants in equation (2.68) are less important than errors in free energies of cluster formation, which are exponentiated, it is evident from Figure 6.9 that the CNT model for condensation rates can produce a substantial error in calculating nucleation rates.

The reason that the molecular dynamics calculations of Li and Truhlar (2008) predict higher rate constants than hard-sphere kinetic theory is that they account for the attractive interatomic potential of aluminum. In effect, attractive forces increase the collision diameter in equation (3.38), and thus the collision rate in equation (3.44). Note

[9] Note that Li and Truhlar (2008) neglect to divide the rate constant for dimerization by two.

that this effect is different from the effect of the nonunity sticking coefficient, which is accounted for in the Li and Truhlar (2008) calculations as a reaction probability.

6.6   Use of Atomistic Data in Calculations of Steady-State Nucleation

6.6.1   Extrapolating Atomistic Values to Large Cluster Sizes

While using atomistic data on free energies of cluster formation in equation (2.68) is in principle straightforward, the examples discussed in the previous section illustrate that there are practical limitations. Most importantly, as such data can only be expected to extend up to a finite cluster size, one cannot know a priori whether the range of available data encompasses the critical size for a given temperature and saturation ratio. For example, considering Figure 6.7, there is a clear maximum in the ΔG_n data for aluminum at 2,000 K and a saturation ratio of 20 at a cluster size of 55. However, as the atomistic data extend only up to size 60, one does not know whether a higher maximum exists for a cluster size larger than 60. In lieu of such data, one could envision a hybrid approach, which makes use of atomistic data at small cluster sizes, up to the largest size for which such data are available, and then uses CNT for the stepwise free energy changes at larger sizes. This approach is in principle superior to CNT, even if the critical size is larger than that for which atomistic data are available, as ΔG_n is the sum of all the stepwise free energy changes up to size n.

Suppose that atomistic data are available up to some size n_0. Then, from equations (2.26) and (3.27), the extrapolated free energy of formation at any size larger than n_0 could then be estimated as

$$\frac{\Delta G_n}{k_B T} = \frac{\Delta G_{n_0}}{k_B T} + \Theta \left(n^{2/3} - n_0^{2/3} \right) - (n - n_0) \ln S. \qquad (6.16)$$

Figure 6.10 shows such an example, in which the atomistic data of Li et al. for ΔG_n for aluminum clusters at 2,000 K and a saturation ratio of 20 are extended beyond size 60 using CNT for the stepwise free energy change.

Even for substances for which atomistic calculations of cluster free energies exist, typically they extend only up to a size n_0 that is much smaller than Li et al.'s calculations for aluminum clusters. In such cases, equation (6.16), together with equation (2.68), could represent an approach to calculating steady-state nucleation rates that at least allows one to improve CNT's predictions by facilitating the utilization of atomistic data at small cluster sizes.

6.6.2   Toward Construction of a "Master Table" of Standard Free Energies of Cluster Formation

With the long-range goal of constructing tables of standard free energies of formation of clusters of various substances, one could envision a hybrid approach in which atomistic values from computational chemistry were used at small cluster sizes, up to the size for which they were available at given temperature, while experimental measurements of

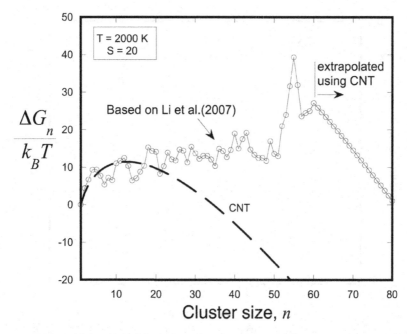

Figure 6.10 Free energies of formation of aluminum clusters at 2,000 K and $S = 20$, based on the atomistic calculations of Li et al. (2007), extrapolated beyond size 60 by assuming the stepwise free energy change from CNT. Adapted from Girshick et al. (2009) with the permission of AIP Publishing.

nucleation rates at the same temperature were used to infer both the critical size n^* and the value of the free energy of cluster formation at the critical size, ΔG^*.

As an example, consider water at 250 K, for which Figure 6.3 shows the values of $\Delta G_n(p_s)$ based on the computational chemistry calculations of ΔG_n^0 by Temelso et al. (2011). Figure 6.11 shows the measured water nucleation rates at 250 K reported by Wölk and Strey (2001). The data consist of 64 measurements over the range of saturation ratio $S = 7.98$ to 10.60. A least-squares power-law fit to these data yields the straight line shown in the figure, corresponding to $J \propto S^{32.515}$. Using the nucleation theorem, in the form of equation (2.87), and rounding to the nearest integer, one obtains $n^* = 32$.

As the critical size is not expected to be constant over this range of saturation ratio, let us assume that $n^* = 32$ applies at $S = 9.20$, which is the geometric mean of 7.98 and 10.60. At $S = 9.20$, again from a least-squares fit to the Wölk and Strey data, one infers a steady-state nucleation rate of $J = 3.81 \times 10^7 \text{cm}^{-3} \cdot \text{s}^{-1}$.

This information can then be used to solve the inverse problem: Given measurements of nucleation rate at given temperature and saturation ratio, what value of ΔG^* would be implied?[10] In this case, we can use the experimental data to estimate the value of ΔG_{32} for water at 250 K and a saturation ratio of 9.2. This could be done using equation

---

[10] The inverse problem involves inferring both n^* and ΔG^* from the measurements of nucleation rate.

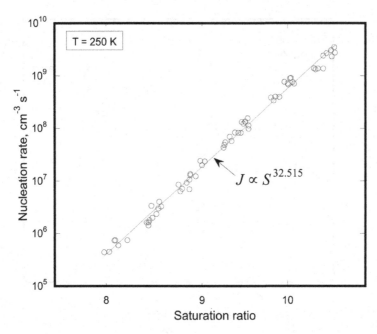

Figure 6.11 Measured water nucleation rates at 250 K reported by Wölk and Strey (2001). The straight line is a linear least-squares fit to the data for $\ln J$ vs. $\ln S$.

(3.91), or, equivalently, equation (3.94), or by means of the "one-term approximation," equation (2.84).

Using the latter approach, and solving equation (2.84) for $\Delta G^*(p_1)$ with the constant C assumed to equal unity, gives

$$\frac{\Delta G^*}{k_B T} \approx \ln\left(\frac{k_{n^*+1} N_1^2}{J}\right). \tag{6.17}$$

In Figure 6.12, a data point based on this exercise has been added to the atomistic values of $\Delta G_n(250\ \text{K}, S = 9.2)$, based on the computational chemistry calculations of Temelso et al. (2011). The value of ΔG_{32} shown in this figure was obtained by means of equation (6.17), with the forward rate constant k_{n^*+1} based on equation (3.47).

Within the uncertainty in the measurement of J, the one-term approximation represents an upper bound on the value of ΔG^*, as if other terms contribute significantly to the summation in equation (2.81), then ΔG^* would be lower, for a given value of J. However, in most cases this error can be expected to be relatively small, both because the factor C in equation (2.84) is not likely to be much greater than unity and because it is in the argument of the logarithm. Moreover, the fact that the nucleation rate in equation (6.17) appears only through its logarithm implies that relatively large errors in the measurement of J would cause relatively small errors in the inferred value of ΔG^*, especially if $\Delta G^* \gg k_B T$, as is usually the case.

Of course, this exercise in itself cannot be expected to yield accurate predictions of nucleation rates at temperatures and saturation ratios that differ from the experimental conditions. However, for some substances, notably including water, experimental

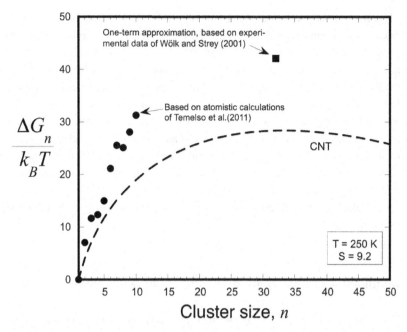

$$\frac{\Delta G_n}{k_B T}$$

One-term approximation, based on experimental data of Wölk and Strey (2001)

Based on atomistic calculations
of Temelso et al.(2011)

CNT

T = 250 K
S = 9.2

Cluster size, n

Figure 6.12 Gibbs free energy of forming $(H_2O)_n$ clusters from monomer vapor at 250 K and $S = 9.2$. Values up to $n = 10$ are based on the calculations of Temelso et al. (2011). The value at $n = 32$ uses one-term approximation of nucleation rate, equation (2.84), based on the measurements of Wölk and Strey (2001).

measurements of nucleation rates have been made at many temperatures and saturation ratios. One can thus envision adding such data, as they accumulate both from computational chemistry and from experimental studies, to a master table of standard free energies of cluster formation, $\Delta G_n^0(T)$, for clusters of each size, up to the size for which such data are available.

Even for the limited set of substances for which, to date, both computational chemistry calculations and experimental studies have been reported at the same temperature, considerable gaps in cluster size space typically exist between the largest cluster studied by computational chemistry and the smallest cluster whose value of ΔG^* can be inferred from experiments. Thus, as a placeholder until more data become available, one might bridge these gaps with some sort of interpolation within each graph of the form of Figure 6.12, while, for clusters larger than the largest size for which data are available, one could use an $n^{2/3}$ extrapolation, as in Figure 6.10.

The resulting "master table" of $\Delta G_n^0(T)$ would not yet be expected to yield accurate predictions of nucleation rates for conditions outside of those that have already been studied experimentally. For one thing, the data, both from computational chemistry and from experimental measurements of nucleation rates, involve considerable uncertainties. For example, Afzalifar et al. (2022) compare the results of several different quantum chemistry simulation methods in calculating the stepwise free energy change, $\Delta G_{n-1,n}^0$, for n up to 14, for water over temperatures ranging from around 200 to 320 K. The range of these predictions at each temperature is considerable.

Moreover, as the error bars in Figures 3.20–3.22 illustrate, measurements of nucleation rates involve considerable uncertainties. Nevertheless, this hybrid approach should in principle be much more accurate than the crude model of CNT, and both computation and experiment can be expected to become more accurate over time.

Given the practical computational constraints on conducting accurate quantum chemistry studies of clusters of most substances beyond rather small sizes, and the enormous increase in computing power required to expand to larger sizes, this project might seem fanciful. However, over the next few decades one can expect dramatic increases in computing power, perhaps through quantum computing. Together with anticipated advances in artificial intelligence, what seems far-fetched now may be realistic by mid-century or sooner. In that case, one may have "JANAF"-like tables of $\Delta G_n^0(T)$ up to a useful size, for example, $n \approx 100$, and over useful ranges of temperature, for many substances.

6.7   Atomistic Approaches to Multicomponent Nucleation

A number of recent studies have used computational chemistry methods to estimate free energies of cluster formation in multicomponent nucleation, particularly for chemical systems of importance in atmospheric nucleation, and this is presently an active area of research. The review by Elm et al. (2020) lists 15 such studies of binary sulfuric acid–water clusters; 36 studies of ternary clusters consisting of sulfuric acid, a base, and water; 34 studies of binary clusters consisting of sulfuric acid together with various bases; 8 studies of binary clusters of methanesulfonic acid and iodic acid; and 10 studies of multicomponent ionic clusters that include sulfuric acid.

For one example, Rasmussen et al. (2020) used high-level computational chemistry to study binary clusters consisting of two to four sulfuric acid molecules and up to five water molecules. Figure 6.13 shows the structures that for each stoichiometry in their calculations had the lowest free energy at 298.15 K and 1 atm. Table 6.1 shows their calculated standard stepwise free energy changes, $\Delta G_{m;n-1,n}^0$, for adding one water molecule to such a cluster, that is, for the reaction

$$(H_2SO_4)_m(H_2O)_{n-1} + H_2O \rightarrow (H_2SO_4)_m(H_2O)_n, \tag{6.17}$$

again at 298.15 K and 1 atm, for $m = 2 - 4$ and $n = 1 - 5$, together with the results of Kildgaard et al. (2018) for the case $m = 1$.

These calculations are extremely computation intensive. Even at these small sizes, Rasmussen et al. (2020) identified a total of 1,145 unique cluster structures. Moreover, the number of possible structures increases dramatically as cluster size increases.

The generation of such data, even at small sizes, represents an important advance in understanding at least the initial pathways to nucleation in multicomponent systems. It can be anticipated that the database of atomistic free energy data for such systems will continue to grow. Nevertheless, it remains an enormous computational challenge to advance this field to sizes that constitute saddle points in free energy space. It is not presently clear whether this can realistically be achieved in the foreseeable future.

Table 6.1 Stepwise standard free energy change, $\Delta G^0_{m;\, n-1,n}$ (kcal $\cdot$ mol^{-1}) for adding one water molecule to an $(H_2O)_{n-1}(H_2SO_4)_m$ cluster, reaction (6.17), at 298.15 K and 1 atm.

n\m	1	2	3	4
1	−1.7	−2.7	−2.1	−2.8
2	−1.0	−2.5	−2.4	−5.2
3	−1.1	−0.7	−2.2	−2.1
4	−1.1	−0.9	−2.4	−2.4
5	+1.6	−1.5	−3.2	−1.8

Results for $m = 1$ are from Kildgaard et al. (2018), while results for $m = 2$–4 are from Rasmussen et al. (2020). Adapted with permission from Rasmussen et al. (2020). Copyright 2020 American Chemical Society.

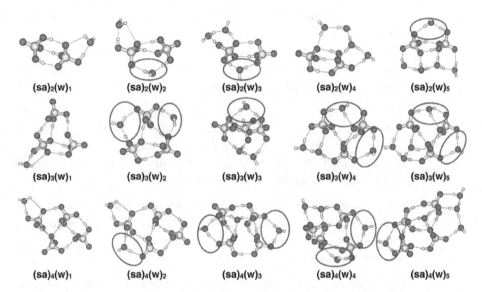

Figure 6.13 Molecular structure of binary clusters containing 2–4 sulfuric acid (sa) molecules and 1–5 water (w) molecules, calculated by Rasmussen et al. (2020) using quantum computational chemistry at 298.15 K and 1 atm. Each cluster shown is the geometry with the lowest free energy configuration for a given stoichiometry. Circles indicate a recurring hydrogen-bonding pattern found in many of the clusters. Reprinted with permission from Rasmussen et al. (2020). Copyright 2020 American Chemical Society.

As noted in Section 4.4.1, the dominant cluster flow path in multicomponent composition space involves a combination of thermodynamic and kinetic factors. In classical theory, the kinetic factors arise mainly because the mole fractions of the participating species are generally different in the gas phase than in the liquid phase. In any atomistic theory, the rate constants for all stepwise reactions are dependent on the detailed structure of the clusters involved. Thus atomistic calculations of rate parameters for cluster growth, and not just the thermodynamic properties, are also important for advancing understanding of multicomponent nucleation.

Homework Problems

6.1 Consider the paper by Temelso et al. (2011), which reports the computational chemistry calculations of Gibbs free energy changes for $(H_2O)_n$ clusters up to $n = 10$, at temperatures ranging from 0 to 373.15 K. Table 4 of the paper gives ΔG_n^0, and Table 5 gives $\Delta G_{n-1,n}^0$, in units of kcal · mol^{-1}, with $p^0 = 1$ atm. (The paper uses a somewhat different notation, but this is what these two tables show in terms of the notation used in this text.)

Based on this information, prepare the following two graphs, both for a temperature of 250 K:

(a) $\Delta G_{n-1,n}(p_s)/k_BT$ vs. n, for n from 2 to 10

(b) $\Delta G_n(p_s)/k_BT$ vs. n, for n from 1 to 10

On both graphs, show also the values predicted by self-consistent CNT, as a continuous curve.

Based on graph (b), do you expect the steady-state nucleation rate predicted by the atomistic calculations to be greater or less than that predicted by CNT, for arbitrary saturation ratio? Explain your answer.

6.2 The standard Gibbs free energies of formation (thermochemical definition, $\Delta_f G_n^0$) of the gas-phase water monomer and dimer, $(H_2O)_2$, evaluated at 300 K, 1 bar, equal -228.504 kJ · mol^{-1} and -449.446 kJ · mol^{-1}, respectively (Ruscic 2013). Based on this information, determine the Gibbs free energy of formation from the monomer vapor, ΔG_n, of the dimer, $(H_2O)_2$, at 300 K, 1 bar.

6.3 Brus et al. (2008a) used a thermal diffusion cloud chamber to measure water nucleation rates at temperatures ranging from 290 to 320 K. Their results from each of the two series of measurements are tabulated in Table 2 in their paper. Combining both sets of measurements made at 300 K:

(a) Prepare a graph of nucleation rate versus saturation ratio, on a log–log scale, and show a least-squares power-law fit to these data.

(b) Assuming that this linear fit is most accurate at the midpoint of the range of $\ln S$ on your graph (i.e., at the geometric mean of the minimum and maximum values of S considered), determine the corresponding value of the critical cluster size at that saturation ratio.

(c) Using the same approach as described in Section 6.6.2, determine the value of $\Delta G^*/k_BT$, at that saturation ratio.

6.4 Based on the data in Table 6.1, what is the change in Gibbs free energy for the reaction

$$(H_2SO_4)_4 + 5(H_2O) \rightarrow (H_2SO_4)_4(H_2O)_5,$$

at 298.15 K, 1 atm?

7 Transient Nucleation

The assumption of the existence of steady-state nucleation is valid in many cases but not all. In this chapter, we consider the conditions required for steady state to exist, the time required to reach steady state, and the modeling of nucleation under conditions where steady state does not exist.

7.1   Assumptions Underlying the Steady-State Nucleation Approximation

The significance of the assumption of steady state derives from equation (2.35),

$$\frac{dN_n}{dt} = J_n - J_{n+1}, \ n \geq 2, \tag{2.35}$$

which relates the nucleation currents at each cluster size to the rate of change in the n-mer number density. This equation implies that the existence of steady state involves two conditions simultaneously: first, that the cluster size distribution N_n is steady in time,

$$\frac{dN_n}{dt} = 0, \ n \geq 2, \tag{2.41}$$

and second, that the nucleation currents J_n at each cluster size are identical to each other and thus can be equated to the steady-state nucleation rate J_{SS}, where we write the subscript "SS" for "steady state," so as to distinguish explicitly between steady-state and transient nucleation.

As noted in Section 2.5, several conditions must be satisfied for the above two equations to hold. First, the validity of equation (2.35) rests on the assumption that there are no n-mer sources or sinks due to the transport of n-mers across system boundaries and also that the rates of cluster–cluster reactions are negligible compared to those of monomer condensation/evaporation events. Second, the validity of equation (2.41) requires that the cluster size distribution N_n adjusts to changes in the nucleation forcing conditions on timescales that are much shorter than the timescales for changes in those conditions.

To clarify the relation of these statements to the physical conditions of a system, we first derive a general expression for the rate of change of N_n at a given location in space, where all of the above assumptions are relaxed.

7.2   General Expression for the Rate of Change in Cluster Number Densities

Consider the system defined by the infinitesimal region of space of volume δV, at location $\mathbf{r}$, depicted in Figure 7.1. With respect to an x-y-z Cartesian coordinate system, $\mathbf{r}$ is located at (x,y,z), and the dimensions of δV can be denoted by δx, δy, and δz. We wish to develop an expression for the rate of change of the n-mer number density $N_n(\mathbf{r}, t)$ in δV.

Clusters of size n can be transported into and out of δV by several effects. In addition, n-mers can be generated or destroyed inside δV by reactions, including both monomer condensation/evaporation "reactions" and other chemical reactions.

Figure 7.2 illustrates the terms for which one must account in formulating an expression for the rate of change in the n-mer number density, $\partial N_n(\mathbf{r}, t)/\partial t$ within δV. For simplicity, these terms are depicted in a two-dimensional (x-y) space.

If the bulk fluid velocity is denoted by vector $\mathbf{u}$, then the flux of n-mers $(\mathrm{m}^{-2} \cdot \mathrm{s}^{-1})$ convected with this flow into δV across the x-pointing face at location x is given by

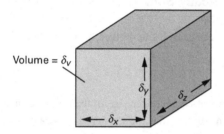

Figure 7.1 Infinitesimal control volume for the derivation of n-mer population balance equation.

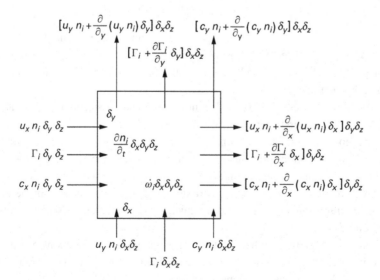

Figure 7.2 Terms that contribute to n-mer population balance equation.

$u_x N_n$, where u_x is the component of $\mathbf{u}$ in the x-direction. At the opposite face, located at $x + \delta x$, n-mers are convected out of δV with a flux that can be written as

$$u_x N_n + \frac{\partial(u_x N_n)}{\partial x} \delta x.$$

Noting that the area of each of the x-pointing faces is given by $\delta_y \delta_z$, the net rate of inflow (s^{-1}) of n-mers into δV due to x-directed convection is thus given by

$$\left\{ u_x N_n - \left[u_x N_n + \frac{\partial(u_x N_n)}{\partial x} \delta x \right] \right\} \delta_y \delta_z = -\frac{\partial(u_x N_n)}{\partial x} \delta V. \tag{7.1}$$

Analogous expressions are readily obtained for the net inflow rate of n-mers into δV with convection in the y- and z-directions. The net convective inflow rate into δV is thus given by

$$-\left[\frac{\partial(u_x N_n)}{\partial x} + \frac{\partial(u_y N_n)}{\partial y} + \frac{\partial(u_z N_n)}{\partial z} \right] \delta V = -\nabla \cdot (\mathbf{u} N_n) \delta V, \tag{7.2}$$

where we have introduced the notation $\nabla \cdot \mathbf{a}$ for the divergence of a vector $\mathbf{a}$.

If there are spatial gradients in N_n, then these gradients drive transport of n-mers by Brownian diffusion across the faces of δV, as given by Fick's law, which can be written in vector notation as

$$\mathbf{\Gamma}_n = -D_n \nabla N_n, \tag{7.3}$$

where $\mathbf{\Gamma}_n$ is the diffusive flux $(\mathrm{m}^{-2} \cdot \mathrm{s}^{-1})$ of n-mers, D_n is the diffusion coefficient $(\mathrm{m}^{-2} \cdot \mathrm{s}^{-1})$ of n-mers in the gas that contains them, and ∇N_n is the spatial gradient in the n-mer number density. The negative sign in equation (7.3) implies that $\mathbf{\Gamma}_n$ points in the direction opposite to the gradient ∇N_n. Following the same derivation as for equation (7.2), and utilizing Fick's law, one obtains that the net inflow rate of n-mers into δV due to diffusion is given by

$$-\nabla \cdot \mathbf{\Gamma}_n \delta V = (\nabla \cdot D_n \nabla N_n) \delta V. \tag{7.4}$$

In addition to convection and diffusion, there are several other effects than can transport clusters across the faces of δV. These include gravity, electrostatic forces, thermophoresis, and other "phoretic" effects such as photophoresis and diffusiophoresis.

Due to their small mass, gravity is unlikely to be a significant effect for clusters in the size range of interest for nucleation.[1] For charged clusters in an electric field, electrostatic forces can be important. For nonisothermal systems, thermophoresis – transport of clusters or particles from hotter toward colder regions, due to the temperature gradient – can be an important effect, especially in cases of steep temperature gradients.

For our purposes, let us group together all of these transport effects, other than convection and diffusion, and let $\mathbf{c}_n$ denote the velocity of n-mers due to these combined

---

[1] The gravitational field on planet Earth is assumed here. In a broader astrophysical context, gravity may be significant for small clusters in locations where gravity is much stronger than on Earth.

effects. In some texts, c_n is referred to as the cluster or particle velocity due to "external force fields" (gravitational field, electric field, etc.), although some of the effects, such as thermophoresis and diffusiophoresis, are really due to the effects of gradients (in temperature, species number density, etc.) on cluster or particle transport due to random molecular collisions not external force fields.

In any case, following the same derivation as above for convective transport, one can write that the net inflow rate of n-mers into δV by all transport effects, other than convection and diffusion, is given by

$$-\left[\frac{\partial(c_{n,x}N_n)}{\partial_x} + \frac{\partial(c_{n,y}N_n)}{\partial_y} + \frac{\partial(c_{n,z}N_n)}{\partial_z}\right]\delta V = -\nabla \cdot (\mathbf{c_n}N_n)\delta V. \tag{7.5}$$

As the above term accounts, by definition of $\mathbf{c}_n$, for all effects other than convection and diffusion that can transport n-mers across the faces of δV, it follows that the sum of the terms given by equations (7.2), (7.4), and (7.5) equals the net inflow rate of n-mers into δV due to all effects. Thus one can write

$$[\text{Net inflow rate (s}^{-1}) \text{ of } n\text{-mers into } \delta V] = -[\nabla \cdot (\mathbf{u}N_n) - (\nabla \cdot D_n\nabla N_n) + \nabla \cdot (\mathbf{c_n}N_n)]\delta V. \tag{7.6}$$

In addition to being transported across the faces of δV, n-mers can be generated or destroyed inside δV by reactions. Let $\dot{\omega}_n$ denote the net rate per unit volume of n-mer generation $(\text{m}^{-3} \cdot \text{s}^{-1})$ by reactions. Then the total rate of n-mer generation (s^{-}) by reactions inside δV is given by $\dot{\omega}_n\delta V$. If the only reactions that generate or destroy n-mers are condensation/evaporation reactions of the form of reaction (2.9), then, following the derivation in Section 2.4, $\dot{\omega}_n$ is identical to the net rate of n-mer formation given by equation (2.35), that is, the difference in the nucleation currents for adjacent sizes:

$$\dot{\omega}_n = J_n - J_{n+1}, \ n \geq 2. \tag{7.7}$$

More generally, $\dot{\omega}_n$ can also include "association/dissociation" reactions of the form

$$A_i + A_j \leftrightarrows A_n, \ n = i + j, \tag{7.8}$$

where i and j may each be greater than 1, as well as any other type of reaction that generates or destroys n-mers.

Finally, one can write a population balance for n-mers in δV as follows:

[Rate of change of number of n-mers inside δV]
= [Net inflow rate of n-mers into δV]
+ [Net rate of n-mer generation by reactions inside δV],

where the left-hand side is given by $\frac{\partial N_n}{\partial t}\delta V$.

Utilizing equation (7.6) for the net inflow rate into δV, and dividing all terms by δV, we have

$$\frac{\partial N_n(\mathbf{r}, t)}{\partial t} = -\nabla \cdot (\mathbf{u}N_n) + (\nabla \cdot D_n\nabla N_n) - \nabla \cdot (\mathbf{c_n}N_n) + \dot{\omega}_n. \tag{7.9}$$

This equation provides a general description of the rate of change of the cluster size distribution $N_n(\mathbf{r}, t)$. Solving it requires knowledge of the flow field $\mathbf{u}(\mathbf{r}, t)$, diffusion coefficients D_n, n-mer velocities $\mathbf{c}_n(\mathbf{r}, t)$ due to any pertinent effects such as electrostatic forces, thermophoretic forces, etc., and rates of all pertinent reactions. In turn, many of these terms require knowledge of the thermodynamic state of the gas, for example, the temperature and pressure, as well as external fields such as the electric field. Nucleation involving chemical reactions in addition to the monomer condensation/evaporation reactions so far considered is discussed in Chapter 8. Nucleation in plasmas, discussed in Chapter 9, requires additional equations, because the charging of clusters can cause a strong coupling between the behavior of clusters or particles and the state of the plasma.

For present purposes, we continue to assume that n-mer formation and destruction involve only monomer condensation/evaporation, as described by reaction (2.9). In that case, using equation (7.7) to substitute for $\dot{\omega}_n$, we can write

$$\frac{\partial N_n(\mathbf{r}, t)}{\partial t} = -\nabla \cdot (\mathbf{u}N_n) + (\nabla \cdot D_n \nabla N_n) - \nabla \cdot (\mathbf{c}_n N_n) + J_n - J_{n+1}, \; n \geq 2. \quad (7.10)$$

The reason that equation (7.10) applies only for $n \geq 2$ is that the nucleation current is not defined for monomers. The population balance for monomers is given by equation (7.9), with $n = 1$, where $\dot{\omega}_1$ represents the net rate of monomer generation by all reactions.

We now consider two important special cases of equation (7.10). In the first case, spatial transport effects are absent, and the system is spatially homogeneous but can evolve in time. In the second case, the system is spatially nonuniform and transport effects are present, but the system is steady with respect to time.

7.3 Homogeneous, Time-Dependent Systems without Transport

In the case of a spatially homogeneous system, without bulk fluid motion or other transport effects, the partial derivative on the left-hand side of equation (7.10) can be replaced by an ordinary derivative for $N_n(t)$, and equation (7.10) reduces to equation (2.35).

We now pose the following problem. Suppose that a supersaturated vapor, consisting only of monomers, exists at time zero with a specified saturation ratio. We assume cluster growth by monomer addition only. For times $t > 0$, clusters begin to form. Let the saturation ratio and temperature be held constant. How long will it take for the nucleation currents to reach steady state, and how do the cluster number densities and nucleation currents behave during the approach to steady state?

To address this problem, we consider the coupled system of equations (2.34) and (2.35) for the cluster number densities N_n and nucleation currents J_n, from $n = 2$ up to some large size M:

$$J_n = k_n N_{n-1} N_1 - k_{-n} N_n, \; 2 \leq n \leq M, \quad (7.11)$$

and

$$\frac{dN_n}{dt} = J_n - J_{n+1}, \; 2 \leq n \leq (M - 1). \quad (7.12)$$

As the nucleation rate is equivalent to the rate of formation of critical-size clusters, M should be larger than the critical size n^*, and sufficiently larger than n^* that the choice of M makes negligible difference in the solution for all values of n smaller than n^*.

An accurate solution of this problem requires detailed knowledge of the size-dependent rate constants and free energies of cluster formation, as discussed in Chapter 6. A potentially important factor is the need for a third body to form dimers and possibly larger clusters, which might slow the approach to steady state and could introduce a dependence on total pressure into the time required to reach steady state.

7.3.1   Analysis Based on Classical Theory

Lacking detailed knowledge of rate constants and free energies, one can again obtain physical insight into the problem by utilizing the classical model.

Note that holding the saturation ratio and temperature constant implies that the monomer number density is also held constant, meaning that we ignore the fact that cluster growth depletes the monomer number density. However, at least during the early stages of nucleation, the monomer number density is typically much larger than the number densities of clusters of any size – indeed our neglect of cluster–cluster reactions requires this assumption – so, to first order, the depletion of monomers can be assumed to be negligible during this early phase, which perhaps, depending on conditions, extends beyond the establishment of a steady-state nucleation current. In any case, one of our goals is to determine the time required to reach steady state, for a specified and fixed saturation ratio and temperature.

Using the classical model for the rate constants, with k_n given by equation (3.54) and k_{-n} by equation (3.59), equation (7.11) can be rewritten as

$$J_n = \beta s_{n-1} \left\{ N_{n-1} - \frac{N_s N_n}{N_1} \exp\left\{ \left[n^{2/3} - (n-1)^{2/3} \right] \Theta \right\} \right\}, \quad 2 \leq n \leq M. \quad (7.13)$$

Equations (7.12) and (7.13) can be nondimensionalized by defining a dimensionless nucleation current J_n^*, a dimensionless n-mer number density N_n^*, and a dimensionless time t^*, as follows:

$$J_n^* \equiv \frac{J_n}{\beta s_1 N_1}, \quad (7.14)$$

$$N_n^* \equiv \frac{N_n}{N_1}, \quad (7.15)$$

and

$$t^* \equiv \beta s_1 t. \quad (7.16)$$

Using these definitions, together with equation (3.26) and the definition of saturation ratio, equations (7.13) and (7.12) transform, respectively, to

$$J_n^* = (n-1)^{2/3} \left\{ N_{n-1}^* - \frac{N_n^*}{S} \exp\left\{ \left[n^{2/3} - (n-1)^{2/3} \right] \Theta \right\} \right\}, \quad 2 \leq n \leq M, \quad (7.17)$$

and

$$\frac{dN_n^*}{dt^*} = J_n^* - J_{n+1}^*, \; 2 \leq n \leq (M-1). \tag{7.18}$$

Equations (7.17) and (7.18) comprise two coupled equations for each value of n, for the dimensionless nucleation current $J_n^*(t^*)$ and dimensionless n-mer number density $N_n^*(t^*)$. For the problem posed at the beginning of this section, the initial condition can be written as

$$N_1^*(0) = 1, \tag{7.19}$$

$$N_n^*(0) = 0, \; 2 \leq n \leq M. \tag{7.20}$$

In addition, for times $t^* > 0$, boundary conditions are required for the values of N_n^* for $n = 1$ and $n = M$.

A boundary condition for the monomer number density is required because the calculation of J_2^* in equation (7.17) requires knowledge of N_1^*. By equation (7.15), this boundary condition is simply

$$N_1^* = 1, \; t^* > 0. \tag{7.21}$$

A boundary condition for N_M^* is required because truncation at size M prevents M-mers from having a sink via condensation to form $(M+1)$-mers. This would cause an unrealistic buildup in the M-mer number density, which in turn would affect the n-mer number densities at smaller sizes, through equation (7.18). As M is large, the number density of M-mers can be assumed to be much smaller than that of monomers, that is,

$$N_M^* \ll 1. \tag{7.22}$$

Therefore, a reasonable approximation for this boundary condition is

$$N_M^* = 0, \; t^* > 0. \tag{7.23}$$

The test of whether this choice of boundary condition is suitable is that the value chosen for M should have negligible effect on the results, as long as M is sufficiently larger than the critical size.

Equations (7.17) and (7.18), with auxiliary conditions (7.19)–(7.21) and (7.23), can be solved numerically for given values of Θ and S, and for a value of M that is significantly greater than the critical size given by equation (3.32). One can first solve for all the J_n^*'s at time zero and use these to calculate the values of dN_n^*/dt^*, and then march forward in time to integrate equation (7.18), thus obtaining the values of all the cluster number densities at the end of each time step.

Obtaining the solution may present some numerical difficulty. This is partly because of the large range of timescales required for clusters of different sizes to reach steady state. More importantly, however, it is because as steady state is approached the forward and reverse rates of the condensation/evaporation reactions become quite close to each other, in which case the nucleation currents equal the relatively small difference between two large numbers, causing potential problems with roundoff error. Abraham (1969), who used the classical model to conduct numerical solutions of transient nucleation, found that double precision was required.

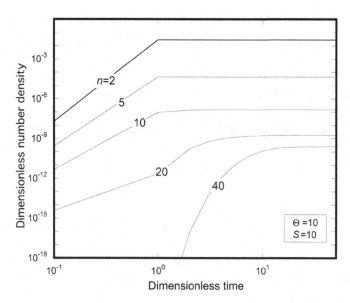

Figure 7.3 Temporal evolution of dimensionless number density for clusters of various sizes, based on CNT, assuming $\Theta = 10$ and $S = 10$. Dimensionless number density and dimensionless time are defined by equations (7.15) and (7.16), respectively.

Figures 7.3 and 7.4 show the results of this calculation for a base case of $\Theta = 10$ and $S = 10$, where M has been set equal to 101. For these conditions, the critical size from equation (3.32) equals 24.3.

As seen in Figure 7.3, the number density of clusters of each size increases with time until reaching steady state. The larger the cluster, the longer it takes to reach this steady state. This makes sense, as the buildup in the density of n-mers must await the buildup in the density of $(n-1)$-mers. For clusters smaller than the critical size, as the number densities increase, the corresponding nucleation currents, shown in Figure 7.4, also increase until reaching a maximum, and then relax to their steady-state values. The existence of these maxima and subsequent relaxation can be explained by the fact that, for each size n, $(n-1)$-mers reach their steady-state densities before $(n+1)$-mers do.

For clusters larger than the critical size, the nucleation currents increase monotonically to their state–state values. The steady-state nucleation currents are seen, as required by equation (7.18), to be the same for all sizes. It is interesting to note that these results for the approach to steady-state nucleation reveal the value of the critical size, even if one did not know it a priori. It is also evident that the existence of the critical size, introduced in Section 3.5.1 based purely on thermodynamic arguments, represents a rate-limiting bottleneck, which forces the nucleation currents for all sizes to eventually equal each other.

7.3.2   The Time Lag for Reaching Steady State

As is evident from Figures 7.3 and 7.4, the establishment of steady-state nucleation requires a certain time lag. Several authors have proposed analytical solutions to the problem of estimating the duration of this time lag (Andres and Boudart 1965; Shi et al.

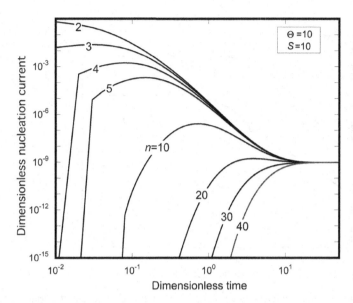

Figure 7.4 Temporal evolution of nucleation current for clusters of various sizes, based on CNT, assuming $\Theta = 10$ and $S = 10$. Dimensionless nucleation current and dimensionless time are defined by equations (7.14) and (7.16), respectively.

1990). The behavior of the transient nucleation current for critical-size clusters, $J_{n*}(t)$, can be approximated by an exponential approach to the steady-state nucleation rate:

$$J_{n*}(t) = J_{SS}\left(1 - e^{-t/\tau}\right), \tag{7.24}$$

where the time lag to reach steady state is defined as the time τ appearing in this equation.

Based on this definition, a dimensionless time lag τ^* can be inferred from our numerical solution as the value t^* for which the dimensionless nucleation current for critical-size clusters equals $(1 - e^{-1})$, or 63.2%, of the dimensionless steady-state nucleation rate. For the case $\Theta = 10$, $S = 10$, taking $n^* = 24$ as the closest integer to the critical size of 24.3, we find $\tau^* = 3.57$.

We can now use equation (7.16) to translate these results to dimensional values of time. For example, for a monomer partial pressure of $p_1 = 100$ Pa, a temperature of $T = 300$ K, and with values of molecular weight and molecular diameter roughly similar to those of water,[2] one finds $\beta s_1 \approx 1.67 \times 10^6\,\text{s}^{-1}$. In this case, a value of 3.57 for τ^* corresponds to a dimensional time lag τ to achieve steady-state nucleation of about 2.14 μs.

Figure 7.5 shows the value of the dimensionless nucleation current, evaluated at the critical size (rounded to the nearest integer), versus dimensionless time for various saturation ratios, again assuming $\Theta = 10$. Higher saturation ratios are seen to cause steady state to be achieved more rapidly. If the partial pressure of the monomer vapor is doubled, causing the saturation ratio to equal 20 instead of 10, the critical size is reduced from 24.3 to 11.0, and our numerical solution indicates that the dimensionless time lag

---

[2] This example is for a hypothetical substance (not water) for which $\Theta(300\text{ K}) = 10$ and $p_s(300\text{ K}) = 10$ Pa.

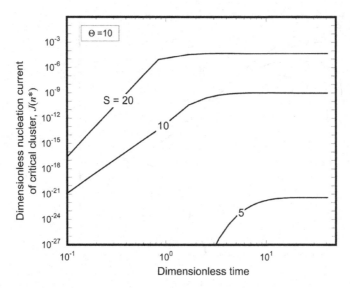

Figure 7.5 Dimensionless nucleation current based on CNT, evaluated at the critical size, versus dimensionless time for various saturation ratios, and for $\Theta = 10$.

τ^* drops from 3.57 to 1.38. Because doubling the partial pressure causes the monomer flux β to double, the effect on the *dimensional* time lag τ is even stronger: It is reduced from 2.14 μs to 0.41 μs. Conversely, if the monomer partial pressure is halved, that is, the saturation ratio equals 5 – corresponding to a critical cluster size of 71.1 – we find $\tau^* = 20.1$ and $\tau = 24.1$ μs.

The time to reach steady state increases for higher values of dimensionless surface tension Θ. For example, assuming a saturation ratio of 10, and for $\Theta = 10$, we find $\tau^* = 3.57$, whereas for $\Theta = 11$, we find $\tau^* = 4.48$.

The effects of S and Θ on the dimensionless time lag τ^* can be understood mainly in terms of their effects on the critical size. The larger the critical size, the longer it takes the "wave" of increase in n-mer number densities to reach the critical size. Furthermore, for a given substance, increases in critical size are associated with decreases in saturation ratio, which cause reduced rates of condensation, thus slowing the forward rate kinetics for each stepwise reaction.

On the other hand, even though the critical size is completely determined by the dimensionless parameters S and Θ, equation (7.16) implies that several parameters affect the conversion to the *dimensional* time lag τ. These include the monomer partial pressure, the temperature, and the mass and surface area of the monomer. Among these, the monomer partial pressure is the most important. This is because, even for given saturation ratios, $p_s(T)$ for different substances varies by many orders of magnitude. For water at 300 K, $p_s = 3.535$ kPa, while for silicon at 298.15 K, the vapor pressure is extrapolated to 9.2×10^{-73} atm (Hultgren 1973), which works out to a number density of about 19 atoms per cubic light year! Evidently, even one stray Si atom in the vapor phase, in a system of terrestrial proportions, at standard temperature and pressure, would constitute a phenomenally supersaturated vapor and would itself constitute a critical nucleus. However, the time lag for reaching steady-state self-nucleation, if such a thing were possible, would be much longer than the estimated lifetime of the universe.

Nevertheless, one finds, for a wide range of scenarios involving appreciable nucleation rates, that the time lag for achieving steady-state nucleation, based on the classical model, usually ranges from a fraction of a microsecond to $\sim$100 µs.

7.3.3 Validity of the Steady-State Assumption for Systems Undergoing Rapid Change

Whether or not transient nucleation can be neglected depends mainly on the timescales for change in the system undergoing nucleation – in particular, the timescale for change in the saturation ratio, which is the most important factor affecting nucleation – relative to the time lag for achieving steady-state nucleation. Under a wide range of conditions, time lags on the order of microseconds are insignificant relative to the timescale for change in conditions of the system, meaning that steady-state nucleation would be an excellent assumption.

In other cases, the validity of the steady-state assumption is less certain. An example is nucleation in supersonic nozzle expansions. Such expansions are an important tool for studying nucleation rates under controlled conditions and are also used as a deliberate method for generating nanoparticles at high rates. In analyzing experimental results, steady-state nucleation is usually assumed.

Supersonic nozzle expansions may achieve cooling rates of $\sim$10^6 to 10^7 K/s, especially in the case of high-temperature, compressible flows. At such a cooling rate, a temperature drop of 10 K requires only 1 to 10 µs. If the mole fraction of condensible monomer is fixed, then a temperature drop of 10 K may cause a significant increase in the saturation ratio. From the preceding analysis, increasing the saturation ratio causes the time lag for steady-state nucleation to decrease. Nevertheless, depending on the substance and on conditions, it is clear that under such high cooling-rate conditions the time lag to achieve steady state may be comparable to the timescale for significant change in the saturation ratio, in which case the system would not be able to adjust rapidly enough to allow steady state to be a valid assumption.

Moreover, the classical model may seriously underestimate the time lag required to achieve steady state nucleation, as discussed in Section 7.3.5.

7.3.4 Effect on Time Lag of Three-Body Reactions at Small Cluster Sizes

As can be seen in Figure 7.3, small clusters reach their steady-state number densities much more rapidly than large clusters. It is thus interesting to ask whether the need for a third body at the smallest cluster sizes, discussed in Section 6.5.2, increases the time lag for reaching steady state, or whether the effect is negligible.

To address this question in an approximate way, we again find a numerical solution of equations (7.17)–(7.21) and (7.23), except that equation (7.17) is modified so as to represent a rate of dimerization that is reduced by an arbitrary factor A. We focus on dimerization as that is the first step in the clustering sequence, so any reduction in its rate is bound to affect the evolution of the nucleation currents and the approach to steady state.

We find that for the case $A = 10^{-2}$, the dimensionless time lag τ^* increases from 3.57 to 6.47, while for case $A = 10^{-4}$, we find $\tau^* = 33.31$. Thus a decrease in the dimerization rate by a factor of 10^4 results in an increase in the time lag by about a factor of 10. For comparison, the silicon dimerization rate seen in Figure 6.8 is reduced by the need for a third body by a factor of $\sim 10^3$. The effect on the nucleation time lag of the third-body requirement for dimerization is thus seen to be substantially damped but is not necessarily negligible. Therefore, this is another factor that can slow the approach to steady-state nucleation in systems at low total pressure.

7.3.5   Analysis Based on Atomistic Data

As discussed in Chapter 6, the use of atomistic data can result in dramatically different cluster free energies, and hence dramatically different steady-state nucleation rates, than those predicted by classical nucleation theory (CNT). These differences also affect the transient nucleation kinetics, including the time lag required to achieve steady-state nucleation.

The transient evolution of nucleation currents and n-mer number densities presented in Figures 7.3–7.5 is based on CNT. What makes this a CNT formulation is equation (7.17), in which the expressions for the nucleation currents derive from the classical expressions for forward and backward rate constants that apply to the sequence of reversible reactions that are assumed to constitute the path to nucleation.

If atomistic data are available for condensation rate constants and cluster free energies, one can utilize these instead of CNT to calculate transient nucleation. Reverting to the description given in Section 2.4, without employing the CNT model, the nucleation currents are given by equation (7.11), while the rate of change of cluster number densities is again given by equation (7.12).

In Section 6.4, two examples are given of substances for which atomistic data on cluster free energies from high-level computational chemistry are available: water, up to clusters of size $n = 10$, and aluminum, up to $n = 60$. Here we present results of transient nucleation calculations for aluminum, which, in addition to having atomistic free energy data up to an exceptionally large size, also has the advantage that atomistic data are also available up to size 60 for condensation rate constants, as shown in Figure 6.9.

Such transient nucleation calculations based on atomistic data can be expected to be highly substance specific. However, even a cursory examination of Figures 6.2 and 6.4, which show the stepwise free energy changes evaluated at $S = 1$ for water and aluminum, respectively, illustrates that the qualitative behavior of the stepwise free energy change based on atomistic data can differ substantially from the smooth, monotonic curve that results from the liquid droplet model. There is no reason to expect that these two substances are exceptional in this regard. And as the stepwise free energy changes mediate the stepwise condensation/evaporation reactions that lie at the heart of the nucleation sequence, it can be anticipated that the transient evolution of the cluster size distribution may be not just quantitatively but also qualitatively different when accounting for atomistic data, compared to the CNT results.

Girshick et al. (2009) conducted transient nucleation calculations based on the computational chemistry calculations for aluminum clusters of Truhlar and coworkers (Li et al. 2007; Li and Truhlar 2008) that are discussed in Chapter 6. Results of these

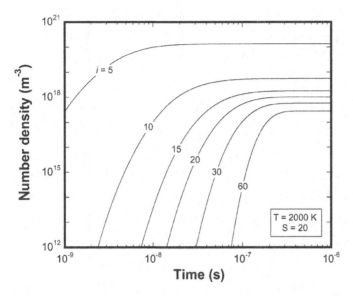

Figure 7.6 Temporal evolution of number densities of aluminum clusters of various sizes i at 2,000 K and a saturation ratio of 20, based on CNT. Reprinted from Girshick et al. (2009) with the permission of AIP Publishing.

calculations were reported for a temperature of 2,000 K and a saturation ratio of 20, for which values of ΔG_n based on the atomistic data are shown in Figure 6.7. As the maximum value of ΔG_n seen in Figure 6.7 occurs at $n = 55$, and as the accuracy of the transient nucleation calculations requires solving equations (7.11) and (7.12) up to a cluster size M that is sufficiently larger than the critical size that the choice of M makes negligible difference in the results, the simulations whose results are shown here set M equal to 100. To facilitate solving equations (7.11) and (7.12) up to size 100, the atomistic data were extrapolated from $n = 60$ to $n = 100$ by assuming that CNT applied beyond size 60 for the stepwise free energy changes, as shown in Figure 6.10, as well as for the condensation rate constants.[3]

It is interesting to note that the numerical difficulty in calculating transient nucleation, discussed in Section 7.3.1, becomes even more severe when using atomistic data than it is in CNT. In the case of aluminum at 2,000 K and $S = 20$, as steady state is approached in the atomistic simulation the value of the nucleation current equals only about one part in 10^{16} of either the forward or backward reaction rates. Girshick et al. (2009) thus found that quadruple precision was required when using the atomistic data, whereas double precision was adequate in the CNT case.

Figures 7.6 and 7.7 show their results for the transient evolution of cluster number densities calculated using either CNT or the atomistic data, respectively.

---

[3] If atomistic data were available beyond size 60, the results might differ from those presented here. The choice of using CNT to extrapolate to size 100 can thus be regarded as a placeholder, lacking better information.

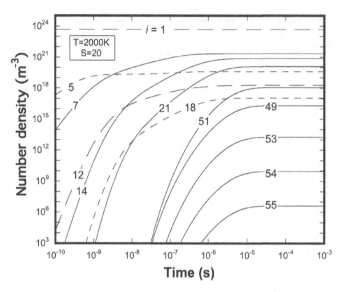

Figure 7.7 Temporal evolution of number densities of aluminum clusters of various sizes i at 2,000 K and a saturation ratio of 20, based on atomistic calculations of Truhlar and coworkers (Li et al. 2007; Li and Truhlar 2008). Reprinted from Girshick et al. (2009) with the permission of AIP Publishing.

In the CNT calculation, Figure 7.6, both the steady-state number density and the time required for clusters of each size to achieve steady-state are seen to vary monotonically with cluster size.

In contrast, the results of the atomistic calculation show remarkably different behavior, as seen in Figure 7.7. The initial conditions fix the monomer number density while setting the number densities of all larger clusters to zero. Therefore, at the earliest times, as clusters of increasing size become successively populated, the cluster number densities must order themselves monotonically with increasing cluster size. In CNT, Figure 7.6, that ordering remains as the system evolves toward steady state, whereas in the atomistic calculation, Figure 7.7, it is seen that the number densities reach steady-state values that are notably *not* ordered by cluster size.

This behavior can be understood by referring to Figure 6.10, which shows that the atomistic values of ΔG_n at 2,000 K and $S = 20$ exhibit a highly multimodal behavior, with multiple local maxima and minima. As discussed in Chapter 6, the local minima in $\Delta G_n(n)$ correspond to magic numbers, whose equilibrium populations are larger than those of neighboring-size clusters. Conversely, the locations of local maxima in a plot of $\Delta G_n(n)$ are referred to as "anti-magic numbers," as their equilibrium populations are smaller than those of clusters of neighboring sizes. This behavior is clearly seen in Figure 7.7. With reference to the plot of $\Delta G_n(n)$ in Figure 6.10, one sees that several of the cluster sizes whose number densities are shown in Figure 7.7 correspond to either magic numbers (local minima in ΔG_n at sizes 7, 14, 21, and 51) or anti-magic numbers (local maxima at sizes 5, 12, 18, 49, and 55). This is most striking for the comparison between clusters of sizes 51 and 55, which comprise the sharpest change in ΔG_n seen in

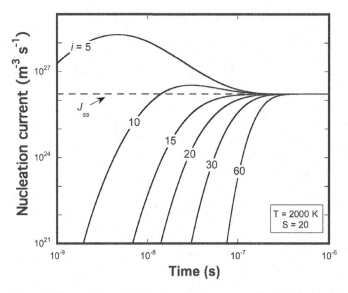

Figure 7.8 Temporal evolution of nucleation currents for aluminum clusters of various sizes i at 2,000 K and a saturation ratio of 20, based on CNT. Line labelled "J_{ss}" is the steady-state nucleation rate given by equation (3.90). Reprinted from Girshick et al. (2009) with the permission of AIP Publishing.

Figure 6.10. At steady state, the calculated number density of clusters of size 51 is greater than that of size 55 by a factor of about 10^{12}.

The temporal evolution of the corresponding nucleation currents for selected cluster sizes calculated by CNT is shown in Figure 7.8. In addition to the nucleation currents for selected cluster sizes, Figure 7.8 also shows the value of the steady-state nucleation rate (dashed line), calculated independently by the CNT analytical expression, equation (3.90). The critical size determined by CNT for this case equals about 12.6. It is seen that the nucleation currents for clusters smaller than this value initially overshoot the steady-state nucleation rate, while clusters larger than x^* do not. Within a few tenths of a microsecond, the nucleation currents for clusters of all sizes asymptotically approach the common steady-state value given by J_{ss}.

As an aside, it is interesting to note that J_{ss}, shown by the dashed line in Figure 7.8, represents the value given by the self-consistent version of CNT. This value is seen to agree excellently with the steady-state nucleation rate inferred from the transient nucleation calculation, based on the convergence of the nucleation currents for different sizes to a common value. The transient calculation of the nucleation currents involves only the stepwise free energy changes, which are identical in either the standard or self-consistent versions of CNT. Yet the standard version gives a value here of the steady-state nucleation rate, from equation (3.107), that is lower by a factor of about 2,000, and so would be in obvious disagreement with the transient calculation. This discrepancy is indicative of the problems resulting from the lack of self-consistency in the standard version.

Figure 7.9 shows the transient nucleation currents for selected cluster sizes calculated using the atomistic data. Also shown is the steady-state nucleation rate (dashed line)

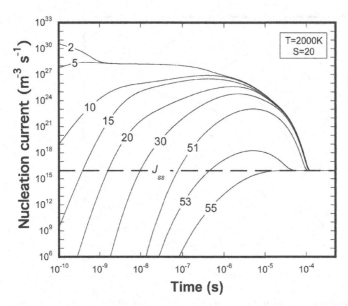

Figure 7.9 Temporal evolution of nucleation currents for aluminum clusters of various sizes at 2,000 K and a saturation ratio of 20, based on atomistic calculations of Truhlar and coworkers (Li et al. 2007; Li and Truhlar 2008). Line labelled "J_{ss}" is the steady-state nucleation rate given by equation (2.80). Reprinted from Girshick et al. (2009) with the permission of AIP Publishing.

calculated from equation (2.80), in which the atomistic data for aluminum clusters have been used. Similar to the behavior seen in the CNT calculation, nucleation currents for clusters smaller than the critical size, which here equals 55, overshoot the steady-state nucleation rate and then converge to the value given by J_{SS}, while nucleation currents for size 55, as well as for larger clusters (not shown), approach the steady-state nucleation rate asymptotically from below. While this behavior is qualitatively similar to that predicted by CNT, an important difference emerges. Namely, the time required to reach steady state is dramatically longer in the atomistic calculation than in CNT: about 10^{-4} s, based on the convergence of the nucleation currents to a common value, compared to only a few tenths of a microsecond for CNT.

The reasons for the much slower approach to steady state in the atomistic calculation compared to CNT are revealed to a large extent by inspection of Figure 6.10, the $\Delta G_n(n)$ curve for this case using either the atomistic data or CNT.

One obvious difference is the much larger critical size in the atomistic case, 55 compared to 12.6. It takes much longer for the buildup in the populations of clusters of each size to reach this larger critical size, and hence it is not surprising that the time required to reach steady state is much longer. Note, however, that there is no apparent reason a priori to expect atomistic data to always indicate a larger critical size than CNT. Rather this effect is presumably substance specific.

On the other hand, another interesting difference in the free energy profiles of the atomistic data versus CNT is seen in Figure 6.10. Namely, while CNT predicts that the

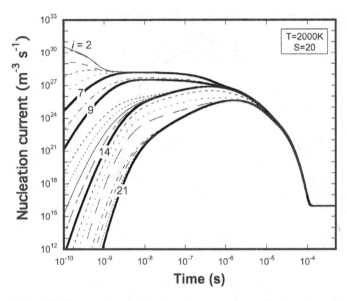

Figure 7.10 Temporal evolution of nucleation currents for aluminum clusters of all sizes $i = 2$ through 21, at 2,000 K and a saturation ratio of 20, based on the atomistic calculations of Truhlar and coworkers (Li et al. 2007; Li and Truhlar 2008). Solid bold lines represent magic numbers. Reprinted from Girshick et al. (2009) with the permission of AIP Publishing.

$\Delta G_n(n)$ curve is monomodal, with a single maximum and no minimum, the atomistic $\Delta G_n(n)$ curve shows the presence of multiple local maxima and minima. Aside from the difference in the critical size, such multimodal free energy profiles can be expected to slow the approach to steady-state nucleation.

The effect of multimodal free energy profiles on the transient nucleation behavior is seen in Figure 7.10, which shows the time-varying nucleation currents for all clusters of sizes 2 through 21, and Figure 7.11, which shows the results for sizes 36 through 55. In both figures, bold solid lines show the nucleation currents for magic numbers, while the bold dashed line in Figure 7.11 shows the nucleation current for size 55, the critical size. For all intermediate sizes shown, the curves are ordered monotonically by increasing cluster size.

In Sections 2.4 and 2.5, where the concept of nucleation current and the distinction between transient and steady-state nucleation are discussed, analogy is made to water flow in a river. Returning to this analogy, consider the flow of water that is released at time zero from a large reservoir and then flows over a riverbed of undulating elevation, with multiple hills separated by basins. The elevation profile of the riverbed is then analogous to the multimodal free energy profile based on atomistic data that is seen in Figure 6.10.

As each basin of the riverbed fills, the currents upstream adjust until a common current is reached, at which time the current downstream of that basin increases, pushing water over the next hill and then on to the next basin, until that basin fills, and so forth. The critical size, corresponding to the highest hill, represents the final rate-limiting bottleneck.

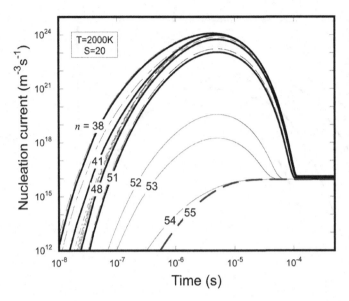

Figure 7.11 Temporal evolution of nucleation currents for aluminum clusters of all sizes $n = 38$ through 55, at 2,000 K and a saturation ratio of 20, based on the atomistic calculations of Truhlar and coworkers (Li et al. 2007; Li and Truhlar 2008). Solid bold lines represent magic numbers. Size 55, the critical size, is an anti-magic number. Reprinted from Girshick et al. (2009) with the permission of AIP Publishing.

Returning to Figures 7.10 and 7.11, this behavior is indeed seen for the entire size spectrum up to the critical size. The nucleation currents for clusters of sizes 2 through 7 (a magic number) become equal to each other after just a few nanoseconds; clusters of sizes 8 and 9 (a magic number) join this equality after a few hundred nanoseconds; clusters of sizes 10 through 14 (a magic number) join at about 1 microsecond; clusters of sizes 15 through 21 (a magic number) join at about 10 microseconds; and by the time the current for size 55 (the critical size, an anti-magic number) has joined this equality, about 100 microseconds have elapsed.

This latter time, about 10^{-4}s, is the time lag required to achieve steady-state nucleation in this case. This time lag, much longer than required by CNT in this case (a few tenths of a microsecond), implies that the assumption of steady-state may often be incorrect. The steady-state assumption is justified by the belief that in most realistic scenarios the time required to reach steady state is much shorter than the timescales for change in the forcing parameters that govern nucleation rates, namely temperature and saturation ratio. However, this assumption derives mainly from the modest literature on transient homogeneous nucleation, which is based mainly on CNT.

As discussed in Section 7.3.3, in situations such as supersonic nozzle expansions the rate of temperature change can equal $10^6 - 10^7$ K/s, with a correspondingly rapid change in the saturation ratio. In this case, if the time lag required to achieve steady state predicted by CNT is only a few tenths of a microsecond, then the assumption of steady state may be borderline but still tenable. On the other hand, an atomistic analysis, such as the example presented here, may in some cases indicate that steady state would not be a valid assumption unless the system conditions were changing much more slowly.

In summary, atomistic analysis indicates that the transient behavior of the nucleation currents for clusters of various sizes is affected by the detailed structure of $\Delta G_n(n)$, and the possibly dramatic increase in the corresponding time lag to achieve steady-state nucleation implies that the assumption of steady state may in many cases be of dubious validity. In such cases, a transient analysis would be necessary.

7.4   Nonuniform Systems with Transport Effects That Are Unchanging with Time

If one considers the case of a nucleating system in which transport effects are present but the system does not change with respect to time, the system can be said to be in "steady state" according to the usual definition. That is, all conditions and quantities are unchanging with respect to time at any given spatial location. In fact, many nucleating systems are well represented by this scenario, for example, steady-flow systems in which a vapor becomes supersaturated as the flow moves downstream. From an Eulerian point of view, which focuses on transfers and rates of change at fixed locations in space, the system may be at steady state, while from a Lagrangian viewpoint, which tracks individual parcels of fluid as they translate through space, key properties, including temperature, saturation ratio, and the cluster size distribution, may be changing with time as one follows a fluid streamline. As discussed in Section 2.5, "steady-state nucleation" requires that the nucleation currents J_n respond rapidly to changes in the nucleation forcing conditions, on timescales that are short relative to the timescales for change in those conditions. The relationships presented throughout this text adopt the Eulerian viewpoint. However, the nucleation currents for each size do not necessarily equal each other, even when the flow is in steady state. Thus, while each of the nucleation currents may be steady with respect to time at a particular location in space, they do not correspond to the special case of steady-state nucleation discussed in Section 2.5.

This would be evident if one adopted a Lagrangian viewpoint. As one follows a parcel of fluid through space, spatial rates of change along a streamline can be transformed to temporal rates of change. The achievement of steady-state nucleation requires that as a fluid parcel moves downstream, the cluster size distribution N_n adjusts on timescales that are very short compared to timescales for changes in nucleation forcing conditions, where time is measured following the flow. Additionally, achievement of steady-state nucleation requires that at each location any net growth or depletion of n-mer number densities due to external sources or sinks – that is, transport effects – must be negligible compared to the growth or depletion by condensation and evaporation. If these conditions are satisfied, then at each location in space the achievement of steady-state nucleation is possible. On the other hand, if these conditions are not satisfied, then steady-state nucleation cannot exist, and calculating nucleation requires solving population balance equations for clusters of each discrete size. The problem is then similar to that involved in the analysis of chemically reacting flows, where concentrations of each chemical species must be calculated at each location. For a simplified one-dimensional analysis that adopts a Lagrangian framework and that explores the effects of convection and diffusion on the possibility to achieve steady-state nucleation; see Suh et al. (2003).

7.5   Transient Nucleation: Summary

When a supersaturated vapor undergoes the series of clustering reactions that lead to particle nucleation, the nucleation currents J_n at each cluster size n initially differ from each other. If conditions such as temperature and saturation ratio are held constant, the existence of a kinetic bottleneck at the critical size causes the system to approach a steady state in which the number densities of clusters of all sizes achieve values that are steady in time, and the nucleation currents J_n for clusters of all sizes become equal to each other, thus constituting the steady-state nucleation rate J. If the nucleation forcing conditions such as temperature and saturation ratio then change, the system adjusts to produce a new steady-state nucleation rate. This adjustment is not instantaneous, but requires finite time. During this transition, the system can be said to undergo transient nucleation.

Most of the literature of single-component homogeneous nucleation assumes the existence of steady-state nucleation. The validity of this assumption depends on how rapidly the size-dependent cluster number densities and nucleation currents adjust to changes in the nucleation forcing conditions.

Numerical simulations of transient nucleation based on CNT indicate that for a wide variety of substances and conditions the time lag required to achieve steady-state nucleation ranges from about 10^{-7} to 10^{-4} s. For many systems, that is sufficiently rapid, relative to the rates of change in temperature and saturation ratio, to make steady-state nucleation a reasonable assumption. However in some cases, such as supersonic nozzle expansions, the temperature and saturation ratio may change so rapidly along a flow streamline that steady-state nucleation is not a valid assumption. Moreover, if instead of CNT one uses atomistic data on cluster free energies and clustering rate constants, the time lag required to achieve steady-state nucleation may be considerably longer than suggested by the classical analysis, rendering the assumption of steady-state nucleation yet more tenuous.

Homework Problems

7.1. In Figure 7.4, which is based on classical theory, it is seen that the nucleation currents for clusters smaller than the critical size reach a maximum and then relax to their steady-state value, whereas for clusters larger than the critical size there is only a monotonic increase. Explain why this is so.

7.2 Write a computer code to solve for transient nucleation for a specific substance and conditions, assuming the classical model, and run a simulation. (Make sure to use double precision!) Compare to the steady-state nucleation rate given by equation (3.90). Evaluate the time lag to achieve steady state for two different saturation ratios.

7.3   Let the time required to reach 99% of the steady-state nucleation rate be denoted τ_{99}. What does τ_{99} equal relative to τ^*, defined in the discussion following equation (7.24)?

8   Chemical Nucleation

In most of the text to this point, nucleation has involved the growth of clusters composed of a given substance or substances, by condensation of vapors of the same substance or substances.[1] However, an important class of nucleation phenomena does not follow this scenario. Instead, nucleation involves the growth of molecular entities ("clusters") by a series of chemical reactions, where none of the chemical species that participate in these reactions is necessarily a supersaturated vapor. Indeed, it is also possible that none of the participating gas-phase species are even of the same chemical substance as the condensed-phase particles that are formed.

8.1   What Is Chemical Nucleation?

An example of chemical nucleation is soot formation in hydrocarbon flames. This is an important nucleation phenomenon, as it is central to particulate formation in automotive engines, coal-fired power plants, and forest fires, with major impacts on health and visibility effects of air pollution, as well as global climate change.

Under the vast majority of practical conditions, soot does not form by condensation of supersaturated monatomic carbon vapor. While monatomic carbon vapor may in some cases exist in nonnegligible concentrations in such systems, and depending on conditions may be supersaturated, it is typically at most a minor contributor to cluster growth. Instead, the carbon atoms added to the growing cluster are contributed by various gaseous hydrocarbon species that undergo chemical reactions with the cluster to grow larger and larger clusters, where none of these growth species are themselves necessarily supersaturated. As the hydrocarbon cluster grows, hydrogen atoms may become preferentially segregated to locations on or near the cluster surface and may then return to the gas phase by either heterogeneous chemical reactions or thermal desorption, finally leaving behind a pure carbon particle or a particle with a carbon core and hydrogen atoms on its surface.

Indeed, the laws of thermodynamics do not require a specific vapor species to be supersaturated for nucleation to occur, only that the overall system is out of equilibrium with respect to the existence of a condensed phase of some substance that can be formed

---

[1] Ion-induced nucleation is a partial exception, in that the ion may be of a different substance than the condensing vapor.

from the elemental constituents of which the system is composed. A suitable kinetic pathway may or may not exist to facilitate formation of that condensed phase on timescales of interest, but that is not the concern of thermodynamics. For example, graphite is the thermodynamically favored form of solid carbon at room temperature and atmospheric pressure, but diamond under such conditions does not convert spontaneously to graphite.

We term nucleation that follows the scenario outlined above "chemical nucleation." Note that the fact that a gas-phase system is chemically reacting and undergoes nucleation does not in itself imply that "chemical nucleation" has occurred, as cluster growth itself might still follow the conventional scenario of nucleation via condensation of a supersaturated vapor. For example, the condensible vapor might be generated by gas-phase chemical reactions, as in many types of engineered aerosol reactors, and may then undergo ordinary self-nucleation. We here reserve the term "chemical nucleation" for the case where cluster growth does not involve the sequential addition of monomers of a supersaturated vapor, but rather chemical reactions among species none of which are themselves necessarily supersaturated.

In describing ordinary self-nucleation, we have referred to the clustering sequence, equations (2.6)–(2.9), as involving "reactions," and have used the formalism and terminology of chemical reactions to describe the behavior of the system, regardless of whether monomer condensation is a physisorption process, with clusters held together by van der Waals forces, or actually involves the breaking and making of chemical bonds. Classical nucleation theory implicitly assumes that clusters, modeled as liquid droplets, grow by physisorption and does not concern itself with chemical reactions. On the other hand, the atomistic approach to self-nucleation discussed in Chapter 6 recognizes that, even in the conventional scenario of condensation of a supersaturated vapor, clusters of specific substances may be structured, chemically bound entities, in which case cluster growth does indeed involve chemical reactions. Nevertheless, for our purposes we do not consider atomistic treatment of self-nucleation to represent chemical nucleation, as it is still driven by condensation of a supersaturated vapor, and the basic cluster growth scenario represented by reactions (2.6)–(2.9) still applies, even though the liquid droplet model of CNT does not apply.

Arguably, chemical nucleation as defined here is important and ubiquitous in nature and technology. Indeed, it is arguably such a common route to gas-phase nucleation – in reality, as opposed to nucleation theory – that it can hardly be termed exceptional. Considering this, it is remarkable that no general theory exists to describe it, though that is perhaps not surprising, considering that study of the simpler case of atomistic treatment of nucleation from a supersaturated vapor is itself in its infancy.

A partial exception exists to the statement that no general theory exists to describe chemical nucleation: the case where cluster growth occurs by a simple repetitive sequence of reversible chemical reactions, so we begin by considering that case. We next discuss two examples of chemical nucleation: soot formation in hydrocarbon combustion and silicon particle formation from thermal decomposition of silane.

Interestingly, carbon and silicon nucleation are also the most studied cases in the literature of particle formation in nonthermal plasmas. We return to the case of silicon nucleation in Chapter 9.

8.2   Nucleation from a Simple Sequence of Reversible Chemical Reactions

In some cases, cluster growth might proceed through a simple repetitive sequence of chemical reactions even though these reactions neither involve any supersaturated vapors nor follow the precise scenario of reactions (2.6)–(2.9). For example, in the 1990s it was proposed by several research groups that nucleation of silicon particles in low-pressure silane plasmas occurs at least partially by the following sequence of reactions (Fridman et al. 1996; Howling et al. 1993; Perrin et al. 1994):

$$SiH_3^- + SiH_4 \rightarrow Si_2H_5^- + H_2 \qquad (8.1)$$

$$Si_2H_5^- + SiH_4 \rightarrow Si_3H_7^- + H_2 \qquad (8.2)$$

$$Si_3H_7^- + SiH_4 \rightarrow Si_4H_9^- + H_2 \qquad (8.3)$$

$$\vdots$$

$$Si_{n-1}H_{2n-1}^- + SiH_4 \rightarrow Si_nH_{2n+1}^- + H_2 \qquad (8.4)$$

Although this particular sequence involves the growth of anion clusters via ion-molecule reactions and occurs in a plasma, for present purposes that is immaterial. What matters is that this is an example of chemical nucleation – cluster growth occurs by chemical reactions, and none of the participating species is itself a supersaturated vapor – that involves a simple repetitive sequence. In this particular example, each reaction involves SiH_4 addition and H_2 elimination and grows the cluster by the addition of one Si atom. Two hydrogen atoms are also added to the cluster in each step, but they might subsequently return to the gas phase by other reactions outside the main clustering sequence. Some types of polymerization might follow a sequence of this general type.

Let us generalize this sequence by rewriting the first step as

$$A_1B_i + A_1B_j \rightarrow A_2B_{i+j-k} + B_k. \qquad (8.5)$$

Here A_1B_i can be termed the "initiating species," A_1B_j the "growth species,"[2] which contributes one A atom to the growing A_nB_m cluster, a role played by SiH_4 in reactions (8.1)–(8.4), and B_k can be termed a "by-product" species of the reaction, just as H_2 can be considered a by-product of reactions (8.1)–(8.4). The subscripts i, j, and k are all nonnegative integers. For simplicity, we suppose that both the initiating species A_1B_i and the growth species A_1B_j contain only one A atom, but it would be straightforward to extend the derivation below to the case where either or both of them contained more than one A atom, or any number of elemental constituents, as long as cluster growth involved a simple repetitive sequence.

Treating reactions of the form of (8.5) as reversible, one can write the reaction sequence as

---

[2] The subscript "1," indicating that the initiating and growth species each contain one A atom, would of course be omitted from usual chemical notation, but is included here for clarity.

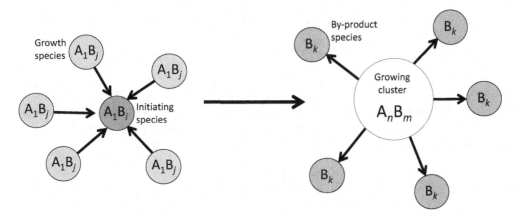

Figure 8.1 Illustration of the overall reaction (8.11) to grow a cluster by a simple repetitive sequence of chemical reactions, (8.6)–(8.9).

$$A_1B_i + A_1B_j \rightleftarrows A_2B_{i+j-k} + B_k \tag{8.6}$$

$$A_2B_{i+j-k} + A_1B_j \rightleftarrows A_3B_{i+2(j-k)} + B_k \tag{8.7}$$

$$A_3B_{i+2(j-k)} + A_1B_j \rightleftarrows A_4B_{i+3(j-k)} + B_k \tag{8.8}$$

$$\vdots$$

$$A_{n-1}B_{i+(n-2)(j-k)} + A_1B_j \rightleftarrows A_nB_m + B_k \tag{8.9}$$

where the nonnegative integer m is given by

$$m = i + (n-1)(j-k). \tag{8.10}$$

Each of reactions (8.6)-(8.9) can be considered a stepwise reaction in that each such reaction contributes (in the forward direction) or removes (in the backward direction) one A atom to/from the cluster. Starting with the initiating species AB_i, the overall reaction to form a cluster containing n A atoms is given by

$$A_1B_i + (n-1)A_1B_j \rightarrow A_nB_m + (n-1)B_k, \tag{8.11}$$

as illustrated in Figure 8.1.

One can derive an expression for the steady-state nucleation rate in this system following an approach that is analogous to that followed in Chapter 2 (Girshick 1997). For economy of notation, let N_n denote the number density of clusters containing n A atoms, k_n the forward rate constant for the stepwise reaction to form such clusters, that is, reaction (8.9), and k_{-n} the backward rate constant for this reaction.

In analogy with the nucleation current defined in Section 2.4, an expression for the nucleation current at each cluster size can then be written as follows:

$$J_2 = k_2N_1N_{AB_j} - k_{-2}N_2N_{B_k} \tag{8.12}$$

$$J_3 = k_3N_2N_{AB_j} - k_{-3}N_3N_{B_k} \tag{8.13}$$

$$J_4 = k_4 N_3 N_{AB_j} - k_{-4} N_4 N_{B_k} \tag{8.14}$$

$$\vdots$$

$$J_n = k_n N_{n-1} N_{AB_j} - k_{-n} N_n N_{B_k} \tag{8.15}$$

Now, at equilibrium the forward and backward reaction rates of each of the reactions (8.6)–(8.9) are equal to each other. With reference to reaction (8.9), at equilibrium one can write

$$\frac{k_n}{k_{-n}} = \left(\frac{N_n N_{B_k}}{N_{n-1} N_{AB_j}} \right)_{eq}, \; n \geq 2. \tag{8.16}$$

The right-hand side of this equation is the equilibrium constant for reaction (8.9) in number density form. As the reactions here happen to be equimolar, this equilibrium constant is equal to the dimensionless equilibrium constant for this stepwise reaction, $K_{n-1,n}$. Thus

$$\frac{k_n}{k_{-n}} = K_{n-1,n}, \; n \geq 2. \tag{8.17}$$

Note that this relation differs from equation (2.20) by the factor $k_B T / p^0$, because the reactions in sequence (2.6)–(2.9) are not equimolar. However, the theory presented here could easily be extended to the nonequimolar case.

Thus the backward rate constants in the expressions for the nucleation currents, reactions (8.12)–(8.15), can be eliminated in favor of the forward rate constants and the equilibrium constants for each reaction, giving

$$J_2 = k_2 N_1 N_{A_1 B_j} - \frac{k_2}{K_{1,2}} N_2 N_{B_k} \tag{8.18}$$

$$J_3 = k_3 N_2 N_{A_1 B_j} - \frac{k_3}{K_{2,3}} N_3 N_{B_k} \tag{8.19}$$

$$J_4 = k_4 N_3 N_{A_1 B_j} - \frac{k_4}{K_{3,4}} N_4 N_{B_k} \tag{8.20}$$

$$\vdots$$

$$J_n = k_4 N_{n-1} N_{A_1 B_j} - \frac{k_n}{K_{n-1,n}} N_n N_{B_k} \tag{8.21}$$

Let us now assume that a steady-state nucleation current is achieved, implying that the nucleation current J becomes independent of cluster size. Then equations (8.18)–(8.21) are all of a form that can be rearranged as

$$\frac{J}{k_n N_{A_1 B_j}} = N_{n-1} - \frac{N_n N_{B_k}}{K_{n-1,n} N_{A_1 B_j}}. \tag{8.22}$$

Let us denote by α the ratio of the number densities of the growth species $A_1 B_j$ and the by-product species B_k:

$$\alpha \equiv \frac{N_{A_1 B_j}}{N_{B_k}},$$ (8.23)

which allows us to write equation (8.22) more compactly as

$$\frac{J}{k_n N_{A_1 B_j}} = N_{n-1} - \frac{N_n}{\alpha K_{n-1,n}}.$$ (8.24)

Following a similar approach as in Section 2.6, we can write this equation for all values of n from $n = 2$ to $n = M$, where M is an arbitrarily large integer:

$$n = 2: \quad \frac{J}{k_2 N_{A_1 B_j}} = N_1 - \frac{N_2}{\alpha K_{1,2}}$$ (8.25)

$$n = 3: \quad \frac{J}{k_3 N_{A_1 B_j}} = N_2 - \frac{N_3}{\alpha K_{2,3}}$$ (8.26)

$$n = 4: \quad \frac{J}{k_4 N_{A_1 B_j}} = N_3 - \frac{N_4}{\alpha K_{3,4}}$$ (8.27)

$$\vdots$$

$$n = M: \quad \frac{J}{k_M N_{A_1 B_j}} = N_{M-1} - \frac{N_M}{\alpha K_{M-1,M}}.$$ (8.28)

We now seek a recursion relation for a function $f(n)$ such that when all these equations are divided by $f(n)$ and then added together all inner terms on the right-hand side cancel, leaving only the first term of the first equation ($n = 2$) and the last term of the final equation ($n = M$). The recursion relation can be found by inspection:

$$f(2) = 1,$$ (8.29)

$$f(n) = f(n-1)\alpha K_{n-2,n-1}, \quad n \geq 3.$$ (8.30)

Carrying out this exercise yields

$$\frac{J}{N_{A_1 B_j}} \sum_{n=2}^{M} \left(k_n \alpha^{n-2} \prod_{i=2}^{n-1} K_{i-1,i} \right)^{-1} = N_1 - \frac{N_M}{\alpha^{M-1} K_{1,2} K_{2,3} \cdots K_{M-1,M}}.$$ (8.31)

The product of the equilibrium constants that appears on the left-hand side of equation (8.31) can now be rewritten in favor of the stepwise standard Gibbs free energy changes of reactions (8.6)–(8.9), using equations (2.21) and (3.24), to give

$$\prod_{i=2}^{n-1} K_{i-1,i} = \exp\left(-\frac{\Delta G_{1,2}^0 + \Delta G_{2,3}^0 + \cdots + \Delta G_{n-2,n-1}^0}{k_B T} \right) = \exp\left(-\frac{\Delta G_{n-1}^0}{k_B T} \right) = K_{n-1},$$ (8.32)

where ΔG_n^0 and K_n are the standard Gibbs free energy of formation of a cluster containing n A atoms via the overall reaction (8.11), and the corresponding equilibrium constant, respectively. Note that $\Delta G_1^0(= 0)$ and $K_1(= 1)$ are defined for the trivial reaction (setting n to 1 in equation (8.11)),

$$A_1B_i \rightleftarrows A_1B_i. \tag{8.33}$$

Thus equation (8.31) can be rewritten as

$$\frac{J}{N_{A_1B_j}} \sum_{n=2}^{M} \left[k_n \alpha^{n-2} \exp\left(-\frac{\Delta G_{n-1}^0}{k_BT}\right) \right]^{-1} = N_1 - \frac{N_M}{\alpha^{M-1}K_M}. \tag{8.34}$$

As the equilibrium constant K_M applies to the overall reaction (8.11), it is defined by

$$K_M = \left(\frac{N_M N_{B_k}^{M-1}}{N_1 N_{A_1B_j}^{M-1}}\right)_{eq} = \left(\frac{N_M/N_1}{\alpha^{M-1}}\right)_{eq}. \tag{8.35}$$

Let α_{eq} denote the value of α that exists in equilibrium at given temperature with any given value (i.e., the actual value) of the ratio N_M/N_1, so that equation (8.35) can be rewritten as

$$K_M = \frac{N_M/N_1}{\alpha_{eq}^{M-1}}. \tag{8.36}$$

Using this expression for the equilibrium constant in equation (8.34) yields

$$\frac{J}{N_{A_1B_j}} \sum_{n=2}^{M} \left[k_n \alpha^{n-2} \exp\left(-\frac{\Delta G_{n-1}^0}{k_BT}\right) \right]^{-1} = N_1 \left[1 - \left(\frac{\alpha_{eq}}{\alpha}\right)^{M-1} \right]. \tag{8.37}$$

Solving for the nucleation rate, and recalling that N_1 is the number density of the initiating species A_1B_i, gives

$$J = N_{A_1B_i} N_{A_1B_j} \left[1 - \left(\frac{\alpha}{\alpha_{eq}}\right)^{-(M-1)} \right] \left\{ \sum_{n=2}^{M} \left[k_n \alpha^{n-2} \exp\left(-\frac{\Delta G_{n-1}^0}{k_BT}\right) \right]^{-1} \right\}^{-1}. \tag{8.38}$$

Comparing this expression to the analogous result for the case of self-nucleation from a supersaturated vapor, equation (2.66), one sees that the expressions are quite similar. In both cases, the steady-state nucleation rate scales on the product of the number densities of the species involved in the first forward reaction in the clustering sequence. In the case of ordinary self-nucleation, this reaction is dimerization by monomer–monomer reactions, hence $J \propto N_1^2$. In the case of chemical nucleation by a simple repetitive sequence of reactions, the nucleation rate is proportional to the product of the number densities of the initiating species, A_1B_i, and the growth species, A_1B_j.

Furthermore, the ratio α/α_{eq} in equation (8.38) is seen to play a role analogous to the saturation ratio in equation (2.66). As M is arbitrarily large, the term $\left[1 - (\alpha/\alpha_{eq})^{-(M-1)} \right]$ is negative if $\alpha/\alpha_{eq} < 1$; zero if $\alpha/\alpha_{eq} = 1$; and approaches unity for large values of M for the case $\alpha/\alpha_{eq} > 1$. This makes sense, as α denotes the ratio of the number density of the growth species to that of the by-product species. Cluster growth consumes the growth species and generates the by-product species. Thus, if $\alpha = \alpha_{eq}$, there is no net tendency for cluster growth to occur, and a value of α exceeding α_{eq} is required for cluster growth. Indeed, with reference to the example of nucleation of silicon particles from silane, it is empirically observed that adding hydrogen, the by-product species, reduces the tendency for nucleation to occur, as it drives cluster growth reactions in the backward direction. Thus the nonequilibrium factor α/α_{eq} can be taken to represent the system's driving force for nucleation via a specific sequential chemical mechanism.

Hence, with the understanding that $\alpha/\alpha_{eq} > 1$ is required for a steady-state nucleation rate to exist, equation (8.38) can be simplified to

$$
J = N_{A_1 B_i} N_{A_1 B_j} \left\{ \sum_{n=2}^{M} \left[k_n \alpha^{n-2} \exp\left(-\frac{\Delta G_{n-1}^0}{k_B T} \right) \right]^{-1} \right\}^{-1}. \tag{8.39}
$$

Finally, shifting indices in the summation,[3] and using the definition of α, we have

$$
J = N_{A_1 B_i} N_{A_1 B_j} \left\{ \sum_{n=1}^{M} \left[k_{n+1} \left(\frac{N_{A_1 B_j}}{N_{B_k}} \right)^{n-1} \exp\left(-\frac{\Delta G_n^0}{k_B T} \right) \right]^{-1} \right\}^{-1}. \tag{8.40}
$$

Thus, for example, for the reaction sequence (8.1)–(8.4), the steady-state nucleation rate is given by

$$
J = N_{SiH_3^-} N_{SiH_4} \left\{ \sum_{n=1}^{M} \left[k_{n+1} \left(\frac{N_{SiH_4}}{N_{H_2}} \right)^{n-1} \exp\left(-\frac{\Delta G_n^0}{k_B T} \right) \right]^{-1} \right\}^{-1}. \tag{8.41}
$$

Of course, to apply equation (8.40) one requires knowledge of the forward rate constants k_n and standard free energies of cluster formation ΔG_n^0 at temperature T up to a large enough value of n to constitute a maximum value of the summand, analogous to a critical size. Here it should again be emphasized that ΔG_n^0 refers specifically to the free energy change for the overall reaction (8.11) and thus differs from the chemical free energy of formation, $\Delta_f G^0$, as discussed in Section 2.3. Most likely ΔG_n^0 will be found via equation (2.26) by advancing the knowledge of the stepwise free energy changes involved in the reaction sequence (8.6)–(8.9).

The formal similarity of equation (8.40) to the result for self-nucleation from a supersaturated vapor, equation (2.68), is obvious. Here, however, it would make little sense to estimate nucleation rates by applying the classical liquid droplet model, as the cluster growth sequence intrinsically involves chemical reactions.

The steady-state nucleation rate expressed in equation (8.40) should presumably be identical to what one would obtain by solving the corresponding set of chemical rate equations up to a suitably large cluster size. Moreover, if steady-state nucleation is not achieved, then the transient behavior of the system can be found only through such a set of rate equations. However, assuming that the steady-state assumption is valid, an

---

[3] As M is arbitrarily large, we retain it as the upper limit of n rather than replacing it by $M - 1$.

advantage of the theory presented here is that it brings chemical nucleation into the same framework as other types of gas-phase nucleation, as discussed in the preceding chapters. Straightforward gas-phase polymerization sequences may present examples of situations where such an approach could be fruitful.

8.3 Examples of Chemical Nucleation

Other than the special case discussed in Section 8.2, the literature of chemical nucleation is specific to the chemical system involved. Below we discuss two examples: nucleation of soot in hydrocarbon combustion and nucleation of silicon particles in thermal decomposition of silane.

8.3.1 Nucleation of Soot in Hydrocarbon Combustion

Soot formation results from incomplete combustion of hydrocarbon fuels in coal-fired electric power plants, internal combustion engines, and forest fires, and is a major component of particulate air pollution, with important consequences for human health (D'Anna 2009; Lighty et al. 2000). Nucleation of soot has been extensively studied and is widely understood to proceed through the formation of polycyclic aromatic hydrocarbons (PAHs), which are molecules consisting of carbon and hydrogen in which the C atoms are bound together in multiple aromatic rings, each usually containing five or six C atoms.[4] The smallest aromatic ring species, consisting of a single six-membered ring, is benzene, C_6H_6 (and, minus an H atom, phenyl, C_6H_5). Benzene is formed by reactions among smaller hydrocarbons, beginning with decomposition of the fuel in a flame. In PAHs, the multiple rings are fused together by sharing a pair of C atoms. This structure confers stability, making PAHs the building blocks for soot formation. The smallest PAH is naphthalene, $C_{10}H_8$, whose structure is equivalent to two benzene rings fused together, and its corresponding radical naphthalenyl, $C_{10}H_7$.

Detailed chemical kinetic models of small hydrocarbon molecules that are the precursors to formation of the first aromatic ring have undergone extensive development and have been continuously refined by comparison with experiments under a wide range of conditions (Blanquart et al. 2009; Wang and Frenklach 1997). These models involve a rich chemistry that considers a large number of species and reactions in the carbon–hydrogen–oxygen system.

Frenklach and coworkers proposed that the key mechanism for carbon addition to hydrocarbon molecules involved abstraction from the molecule of an H atom followed by reaction with acetylene, C_2H_2, known as the "HACA" (for H-abstraction-C_2H_2-addition) mechanism (Frenklach and Wang 1991). They suggested that the first aromatic ring is formed by either of the reactions:

---

[4] The term "soot" is sometimes reserved for nanoparticles larger than a certain size, for example, 10 nm (D'Anna 2009). However, for our purposes, which focus on nucleation, we use "soot" to refer to particles formed by the growth of PAHs during hydrocarbon combustion.

$$n - C_4H_3 + C_2H_2 \rightarrow C_6H_5 \qquad (8.42)$$

or

$$n - C_4H_5 + C_2H_2 \rightarrow C_6H_6 + H. \qquad (8.43)$$

Subsequently Miller and Melius (1992) argued that the first aromatic ring is more likely formed by the recombination of two propargyl radicals, C_3H_3, via either

$$C_3H_3 + C_3H_3 \rightleftarrows C_6H_5 + H \qquad (8.44)$$

or

$$C_3H_3 + C_3H_3 \rightleftarrows C_6H_6. \qquad (8.45)$$

Wang and Frenklach performed numerical simulations that indicated that both channels (i.e., both the $C_4H_x + C_2H_2$ route and propargyl recombination) could be important, depending on conditions (Wang and Frenklach 1997).

Once PAHs are formed, they can grow both by chemical reactions involving small molecules and by PAH coagulation, that is, collision and sticking of PAHs with other (not necessarily the same) PAHs to create PAH dimers, trimers, etc. (Frenklach 2002).

Smooke and coworkers developed a model of soot inception based on chemical reactions of PAHs with smaller molecules (Connelly et al. 2009; Smooke et al. 2004). The initial growth, from naphthalenyl, $C_{10}H_7$, to pyrenyl, $C_{16}H_9$, was assumed to follow a sequence of reactions that is a variant of the HACA mechanism, comprising the addition of three acetylene molecules together with the elimination and abstraction of H atoms, corresponding to the overall reaction

$$C_{10}H_7 + 3C_2H_2 \rightleftarrows C_{16}H_9 + 2H + H_2. \qquad (8.46)$$

They proposed that this reaction sequence is itself the first in a larger sequence that forms large PAHs via the overall reaction

$$C_{10}H_7 + 3nC_2H_2 \rightleftarrows C_{10+6n}H_{7+2n} + 2nH + nH_2. \qquad (8.47)$$

The authors arbitrarily set $n = 21$ to define the minimum size of a "soot particle" (hence, $C_{176}H_{21}$), whose rate of formation was taken to constitute the source term for their model of soot particle dynamics.

In contrast, Pitsch and coworkers emphasized PAH growth by coagulation. In their model, the soot nucleation rate corresponds to the rate of coagulation of PAH dimers with each other (Blanquart and Pitsch 2009), and hence to the formation rate of PAH tetramers. They introduced empirically tuned sticking coefficients for dimerization that scale on the square of the PAH binding energy, which itself scales on the square of the PAH mass. Thus the dimerization sticking coefficients scale on PAH mass to the fourth power.

More recently, a model was proposed that focuses on PAH "chemical dimerization," that is, physical dimerization plus the formation of chemical bonds, as the key step in soot nucleation (Kholghy et al. 2018). A kinetic mechanism was proposed that considers reversible dimerization – physical coagulation, in which dimers are bound together only by van der Waals forces – dehydrogenation from dimers, and dimer bond

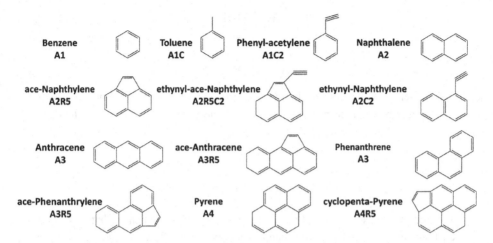

Figure 8.2 Single-ring and polyaromatic hydrocarbons included in the soot nucleation model of Kholghy et al. (2018). Reprinted from Kholghy et al. (2018) with the permission of the Royal Society of Chemistry, conveyed through Copyright Clearance Center, Inc.

formation, in which dimers are bound together by covalent carbon–carbon bonds. The PAHs considered in this model are shown in Figure 8.2.

In all these models of soot nucleation, there is no concept of "critical size" as there is in ordinary self-nucleation. In ordinary self-nucleation, the nucleation rate is equivalent to the rate of formation of clusters of critical size. Similarly, if chemical nucleation can be described by a simple repetitive sequence, as discussed in Section 8.2, one can also identify the critical size as the size associated with the maximum summand in equation (8.41). However, for chemical nucleation in general, it is not evident that the concept of "critical size" is useful, and in any case, no theory presently exists to define it. Rather, numerical models of chemical nucleation typically make an arbitrary assumption regarding the size at which a cluster can be termed a "particle," whose rate of formation can thus be deemed to equal the "nucleation rate."

8.3.2   Nucleation of Silicon Particles in Thermal Decomposition of Silane

Silane, SiH_4, is a widely used process gas in the semiconductor industry for chemical vapor deposition (CVD) of silicon thin films as well as for growth by CVD of silicon rods that are subsequently sliced to produce wafers for microelectronics processing. It is well known that silane is prone to gas-phase nucleation of silicon particles, which in many cases represent undesirable contaminants if they deposit on the material being produced or processed. Alternatively, processes have been developed in which silane is used for the deliberate synthesis of silicon nanoparticles for a variety of applications in optoelectronics, photovoltaics, photonics, and microelectronics.

Two fundamentally different routes for silane decomposition are employed in these industrial applications: either thermally or in a plasma. Here we discuss the case of thermal decomposition. The plasma case is discussed in Chapter 9.

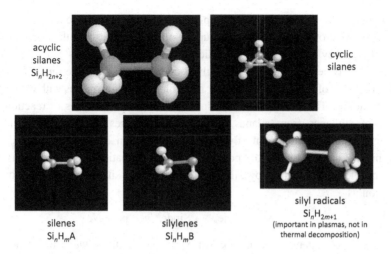

acyclic
silanes
Si_nH_{2n+2}

cyclic
silanes

silenes
Si_nH_mA

silylenes
Si_nH_mB

silyl radicals
Si_nH_{2m+1}
(important in plasmas, not in
thermal decomposition)

Figure 8.3 Main classes of silicon hydrides produced in the dissociation of silane.

As with soot formation in hydrocarbon flames, silicon nucleation from silane is not expected in most cases to proceed via self-nucleation from supersaturated silicon vapor. Rather, under most conditions it occurs by means of reactions among a large number of silicon hydride chemical species. An exception may be high-temperature environments, such as laser pyrolysis and thermal plasmas, where temperatures exceed 2,000 K, which causes the generation of relatively large abundances of bare silicon vapor and suppresses the hydrogenation of growing clusters. However, under environments typically found in silicon CVD, temperatures are more modest – in the range around 1,000 K, depending on pressure – and, while high enough for the dissociation of silane to commence, this results in far lower concentrations of bare Si and far higher tendencies for clusters to be hydrogenated.

Silicon hydrides can be grouped into several classes according to their structure and reactivity, as illustrated in Figure 8.3. Linear, saturated silanes such as silane, disilane (Si_2H_6), and trisilane (Si_3H_8) have the general chemical formula Si_nH_{2n+2}. Silenes (also termed disilenes), including Si_2H_4, Si_3H_6, ... , Si_nH_{2n}, have a double bond between two silicon atoms. (Silenes may also have fewer than $2n$ hydrogen atoms, as long as the number is even.) Silylenes are isomers of silenes that have two nonbonding electrons. In Figure 8.3 and subsequently, silenes and silylenes are distinguished by the suffixes A and B, respectively, in their chemical formulae. Silylenes having at least three silicon atoms can undergo isomerization to form rings of Si atoms, becoming cyclic silanes. As with PAHs, the formation of silicon rings confers stability, making these species important in nucleation.

Removal of one hydrogen atom from SiH_4 creates the silyl radical, SiH_3, and in general silicon hydrides with an odd number of H atoms are classed as silyls. However, silyls are not expected to exist in significant concentrations at the temperatures usually found in thermal decomposition of silicon. This is related to the quite low abundance of monatomic hydrogen at these temperatures and usual pressures, whereas monatomic hydrogen is typically abundant in hydrogen-containing plasmas. Thus, in thermal decomposition silane typically decomposes to SiH_2 and H_2.

Giunta et al. (1990) proposed a simple model of silicon particle nucleation in thermal decomposition of silane, based on grouping silicon hydrides into these generic classes – except that they did not consider cyclic and polycyclic silanes – and assuming that all reactions between two molecules of given classes have the same rate constants, based on knowledge of the rate constants for reactions involving species with only one or two silicon atoms. Their kinetic mechanism considered several types of reactions: elimination of silylenes from silanes, insertion of silylenes into silanes, silene–silylene isomerization, association of silylenes, decomposition of silenes into two silylenes, and nucleation and growth of "powder." Powder formation (i.e., nucleation) itself was assumed to occur mainly by silylene insertion into silanes, written in the authors' notation as

$$\text{syl}_n + \text{ane}_m \rightarrow \text{Sipdr}, \quad n + m > N, \tag{8.48}$$

where n and m denote the number of Si atoms in the silylene or silane molecule, respectively, and N is the number of Si atoms in the largest "non-powder" species considered in their mechanism. Species "Sipdr" thus contains at least $N + 1$ Si atoms and is assumed to grow irreversibly. While analogous to the critical size in conventional self-nucleation, there is no thermodynamic argument advanced to justify the choice of N, which instead is set arbitrarily. Giunta et al. set N equal to 10 in their numerical simulations, based on computational expense considerations together with the observation that choosing a larger value of N would make negligible difference in their results.

A limitation of Giunta et al.'s approach is that, lacking information on the thermodynamic properties of most of the silicon hydrides involved in their mechanism, they specified rate constants only for the forward direction of all reactions. Thus, for example, elimination of silylenes from silanes and insertion of silylenes into silanes were considered separately, each with its own specified rate constant, rather than as a reversible reaction with the reverse rate constant determined by means of an equilibrium constant. The same approach was taken for silene–silylene isomerizations and other types of reactions.

Swihart and Girshick (1999) addressed this limitation by developing a detailed database of the thermodynamic properties of silicon hydrides containing up to 10 silicon atoms. Earlier Katzer et al. (1997) had used an empirically corrected ab initio methodology to calculate the standard enthalpies of formation and entropies of 143 silicon hydride species containing up to five silicon atoms. Swihart and Girshick used these results, together with other unpublished results of Katzer et al., to develop a group additivity approach that allowed them to extend the results to a much larger set of silicon hydrides, containing up to 10 silicon atoms.

The group additivity approach had been used successfully to estimate the thermochemical properties of hydrocarbons (Benson 1976). In this approach, each silicon atom in a silicon hydride molecule is assigned to a group based on its local bonding configuration. Considering the total set of silicon hydrides analyzed by Katzer et al., Swihart and Girshick identified 26 such groups and used a linear least-squares fit to the Katzer et al. data to estimate the contributions of each group to the thermodynamic properties of each species. Additionally, for cyclic species, the contribution of ring strain to enthalpies and entropies was incorporated by including rings of various types among the groups. The results of this exercise are given in Table 8.1 (Swihart and Girshick 1999).

Table 8.1 Group additivity parameters for silicon hydrides, determined by Swihart and Girshick (1999) based on calculations of Katzer et al. (1997).

Group	ΔH_f° 298	S° 298	C_p						
			300	400	500	600	800	1,000	1,500
Si–(H)$_3$(Si)	10.14	33.65	9.68	11.57	13.10	14.38	16.36	17.76	19.68
Si–(H)$_2$(Si)$_2$	9.54	15.74	9.06	10.66	11.83	12.75	14.13	15.08	16.40
Si–(H)(Si)$_3$	7.45	−2.43	8.42	9.64	10.43	11.00	11.80	12.33	13.07
Si–(Si)$_4$	4.40	−22.47	7.94	8.67	9.05	9.25	9.47	9.59	9.75
Si–(H)$_3$(Si$_a$)	10.03	34.06	9.80	11.67	13.15	14.39	16.34	17.72	19.65
Si–(H)$_2$(Si$_a$)(Si)	10.15	15.91	9.21	10.78	11.91	12.80	14.14	15.08	16.39
Si–(H)(Si$_a$)(Si)$_2$	8.99	−2.23	8.75	9.89	10.62	11.14	11.89	12.40	13.10
Si–(Si$_a$)(Si)$_3$	7.90	−21.58	8.37	8.99	9.29	9.45	9.62	9.70	9.81
Si–(H)$_3$(Si$_b$)	10.09	16.33	8.48	10.07	11.30	12.28	13.77	14.81	16.24
Si–(H)$_2$(Si$_b$)(Si)	10.09	−2.52	7.88	9.16	10.00	10.62	11.49	12.08	12.92
Si–(H)(Si$_b$)(Si)$_2$	5.27	−21.03	7.02	7.97	8.47	8.75	9.06	9.25	9.55
Si–(Si$_b$)(Si)$_3$	1.22	−39.03	6.32	6.81	6.91	6.84	6.59	6.39	6.15
Si$_a$–(H)$_2$	33.48	31.87	8.36	9.59	10.50	11.25	12.43	13.29	14.50
Si$_a$–(H)(Si)	31.08	13.81	7.30	8.25	8.89	9.37	10.07	10.55	11.20
Si$_a$–(Si)$_2$	26.77	−3.63	6.08	6.71	7.09	7.33	7.60	7.73	7.85
Si$_b$	66.08	50.31	8.07	8.86	9.53	10.13	11.14	11.86	12.84
C$_3$	36.13	26.65	−2.34	−2.73	−3.04	−3.25	−3.51	−3.65	−3.81
C$_{3a}$	38.96	28.89	−2.58	−2.97	−3.24	−3.42	−3.63	−3.74	−3.86
C$_{3b}$	29.67	28.43	−2.32	−2.75	−3.04	−3.23	−3.45	−3.58	−3.76
C$_4$	16.85	21.89	−3.05	−3.23	−3.38	−3.48	−3.60	−3.69	−3.81
C$_{4a}$	13.07	22.80	−3.31	−3.43	−3.53	−3.61	−3.70	−3.77	−3.86
C$_{4b}$	13.75	23.18	−3.03	−3.22	−3.36	−3.45	−3.56	−3.65	−3.78
C$_5$	4.92	19.62	−4.44	−4.52	−4.57	−4.61	−4.66	−4.72	−4.81
C$_{5a}$	2.47	18.80	−3.72	−3.72	−3.73	−3.74	−3.77	−3.80	−3.86
C$_{5b}$	2.20	19.04	−3.46	−3.54	−3.60	−3.63	−3.69	−3.74	−3.83
C$_6$	0.00	16.80	−2.97	−3.04	−3.09	−3.11	−3.16	−3.24	−3.54

Adapted with the permission of Swihart and Girshick (1999). Copyright 1999 American Chemical Society
Note: Temperatures are in K. Enthalpy contributions are in kcal·mol^{-1}, and entropy and specific heat contributions are in cal·mol^{-1}·K^{-1}.

The essence of group additivity is that the total molecular enthalpy of formation and entropy of a molecule equal the sum of the contributions of the groups that comprise the molecule. An example is shown in Figure 8.4, in this case for the species Si$_5$H$_8$A. Here each of the five silicon atoms is assigned to one of the three groups based on its bonding to other Si and H atoms in the molecule, and additionally the five-membered ring, which in this case includes one Si-Si double bond, is identified as a group. From the database of thermodynamic properties, each of these groups contributes a specified amount to the molecular enthalpy of formation, and these contributions are then summed to give the total enthalpy of formation.

This approach was applied by Swihart and Girshick to generate a table of standard enthalpies and entropies for the large set of silicon hydrides containing from 3 to 10 silicon atoms that is shown in Figure 8.5, which shows their results at a temperature of

Group additivity example: Si₅H₈A

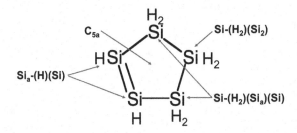

Group	Number	ΔH_f increment (kcal/mol)	Contribution (kcal/mol)
Si-(H₂)(Si₂)	1	9.54	9.54
Si_a-(H)(Si)	2	31.08	62.16
Si-(H₂)(Si_a)(Si)	2	10.15	20.30
C_5a	1	2.47	2.47
		Total ΔH_f	94.5

Figure 8.4 Example of the use of group additivity to estimate total standard enthalpy of formation of silicon hydride, based on Swihart and Girshick (1999).

298.15 K. Additionally, as shown in Table 8.1, values were obtained for the contribution to the specific heat of each of the groups over the temperature range 300–1,000 K, allowing the standard enthalpies and entropies to be extrapolated over that temperature range using the usual ideal gas relations for the dependence of enthalpy and entropy on temperature at given pressure. For molecules of each stoichiometry and each basic type, only the isomer believed to be the most stable was considered.

Based on these thermochemical properties, and the known rate coefficients for reactions involving silicon hydrides with one or two silicon atoms, a kinetic mechanism for silicon hydride clustering was constructed, as shown in Table 8.2. The types of reactions considered included H_2 elimination from silanes, silylene elimination from silanes and silenes, silylene to silene isomerization, ring opening, and the reverse reactions for all of these processes, making use of the equilibrium constant for each reaction as determined by means of the thermochemical database. Additionally, the rate constants for the forward reactions (defined as written in Table 8.2) make use of the thermochemical database by setting the activation energy of each reaction (except for silylene-to-silene isomerization) equal to the enthalpy change ΔH_{rxn} for the reaction, evaluated at 1,000 K, again utilizing the thermochemical database, with the rate constant written in Arrhenius form

$$k = A \exp\left(-\frac{E_a}{RT}\right),$$ (8.49)

E_a being the activation energy.

As with Giunta et al. (1990), the nucleation rate was taken to equal the rate of production of species containing more than 10 silicon atoms. Again, this does not imply a value for a critical size in the same sense as for single-component self-nucleation. Obviously, the kinetic mechanism must be truncated at some cluster size. As seen in Figure 8.5, as the number of silicon atoms increases the number of species that must be

Name	Structure	ΔH_f°	S°	Name	Structure	ΔH_f°	S°	Name	Structure	ΔH_f°	S°	Name	Structure	ΔH_f°	S°
Si_3H_8		29.8	83.0	$Si_6H_{10}B$		114.9	109.3	Si_8H_{16}		72.4	140.3	Si_9H_{12}		128.8	117.9
Si_3H_6A		74.6	79.7	Si_6H_{10}		86.8	101.9	$Si_8H_{14}A$		116.7	136.8	$Si_9H_{10}A$		165.1	118.1
Si_3H_6B		86.3	81.4	Si_6H_8A		120.3	102.0	$Si_8H_{14}B$		125.1	140.4	$Si_9H_{10}B$		177.9	116.6
Si_3H_6		64.8	73.9	Si_6H_8B		138.6	100.9	Si_8H_{14}		82.0	128.8	Si_9H_{10}		141.5	103.5
Si_3H_4A		111.3	72.4	Si_6H_8		118.7	92.2	$Si_8H_{12}A$		118.2	125.5	Si_9H_8A		184.5	100.1
Si_3H_4B		115.9	73.7	Si_6H_6A		154.1	92.4	$Si_8H_{12}B$		134.2	126.0	Si_9H_8B		190.9	100.3
Si_4H_{10}		37.9	98.5	Si_6H_6B		167.8	90.9	Si_8H_{12}		82.7	112.1	$Si_{10}H_{22}$		87.23	188.2
Si_4H_8A		80.3	96.4	Si_6H_6		150.7	82.5	$Si_8H_{10}A$		118.9	108.7	$Si_{10}H_{20}A$		136.9	187.0
Si_4H_8B		91.6	96.6	Si_7H_{16}		62.6	143.4	$Si_8H_{10}B$		129.5	108.3	$Si_{10}H_{20}B$		141.0	186.2
Si_4H_8		55.0	84.9	$Si_7H_{14}A$		108.6	141.7	Si_8H_{10}		131.2	104.5	$Si_{10}H_{20}$		87.5	169.4
Si_4H_6A		95.5	82.2	$Si_7H_{14}B$		115.3	143.5	Si_8H_8A		167.5	104.7	$Si_{10}H_{18}A$		137.6	167.3
Si_4H_6B		109.6	84.2	Si_7H_{14}		65.3	126.7	Si_8H_8B		180.3	103.1	$Si_{10}H_{18}B$		141.4	169.2
Si_5H_{12}		45.0	112.1	$Si_7H_{12}A$		107.3	124.3	Si_8H_8		143.9	90.0	$Si_{10}H_{18}$		91.2	154.7
$Si_5H_{10}A$		87.9	112.4	$Si_7H_{12}B$		119.1	124.8	Si_9H_{20}		80.1	174.6	$Si_{10}H_{16}A$		132.3	152.9
$Si_5H_{10}B$		97.7	112.2	Si_7H_{12}		84.4	115.4	$Si_9H_{18}A$		122.0	174.3	$Si_{10}H_{16}B$		146.1	152.4
Si_5H_{10}		52.6	98.3	$Si_7H_{10}A$		119.2	113.7	$Si_9H_{18}B$		132.9	174.7	$Si_{10}H_{16}$		87.0	135.1
Si_5H_8A		94.5	94.0	$Si_7H_{10}B$		136.2	114.4	Si_9H_{18}		80.4	155.8	$Si_{10}H_{14}A$		130.7	132.5
Si_5H_8B		107.5	95.8	Si_7H_{10}		109.0	103.2	$Si_9H_{16}A$		124.7	152.3	$Si_{10}H_{14}B$		139.2	132.5
Si_5H_8		82.7	85.7	Si_7H_8A		144.4	103.3	$Si_9H_{16}B$		133.2	155.9	$Si_{10}H_{14}$		102.5	126.9
Si_5H_6A		103.8	84.3	Si_7H_8B		158.1	101.8	Si_9H_{16}		86.6	141.7	$Si_{10}H_{12}A$		140.6	123.6
Si_5H_6B		127.6	87.1	Si_7H_8		140.9	93.5	$Si_9H_{14}A$		125.2	139.2	$Si_{10}H_{12}B$		149.3	123.1
Si_6H_{14}		54.5	127.9	Si_7H_6A		177.2	93.7	$Si_9H_{14}B$		138.8	138.9	$Si_{10}H_{12}$		139.1	117.0
$Si_6H_{12}A$		102.9	128.0	Si_7H_6B		190.0	92.1	Si_9H_{14}		87.3	125.0	$Si_{10}H_{10}A$		176.3	117.2
$Si_6H_{12}B$		107.3	128.0	Si_8H_{18}		69.6	157.0	$Si_9H_{12}A$		123.5	121.7	$Si_{10}H_{10}B$		188.5	113.7
Si_6H_{12}		57.2	111.2	$Si_8H_{16}A$		112.9	160.0	$Si_9H_{12}B$		136.8	121.8	$Si_{10}H_{10}$		163.7	104.8
$Si_6H_{10}A$		101.5	107.7	$Si_8H_{16}B$		122.4	157.1								

Figure 8.5 Enthalpy of formation (kcal·mol^{-1}) and standard entropy (cal·mol^{-1}·K^{-1}) at 298.15 K, 1 bar, of silicon hydrides determined by Swihart and Girshick (1999) using group additivity parameters shown in Table 8.1. The structures shown are believed to be the most stable isomer for each stoichiometry and species type. Reprinted with permission from Swihart and Girshick (1999). Copyright 1999 American Chemical Society.

Table 8.2 Reaction types and reactivity rules for silicon hydride clustering mechanism.

Reaction type	Prototypical reactions	General form	Preexponential factor (s^{-1})	Activation energy
H$_2$ elimination from a silane	SiH$_4$ $\leftrightarrow$ SiH$_4$ + H$_2$ Si$_2$H$_6$ $\leftrightarrow$ Si$_2$H$_4$B + H$_2$	Si$_n$H$_{2m}$ $\leftrightarrow$ Si$_n$H$_{2(m-1)}$ + H$_2$	2 $\times$ 10^{15}	$\Delta H_{\text{rxn},1000}$
Silylene elimination from a silane	Si$_2$H$_6$ $\leftrightarrow$ SiH$_2$ + SiH$_4$ Si$_3$H$_8$ $\leftrightarrow$ Si$_2$H$_4$B + SiH$_4$	Si$_n$H$_{2m}$ $\leftrightarrow$ Si$_l$H$_{2k}$B + Si$_{n-l}$H$_{2(m-k)}$	2 $\times$ 10^{15}	$\Delta H_{\text{rxn},1000}$
Silylene elimination from a silene	Si$_3$H$_6$A $\leftrightarrow$ SiH$_2$ + Si$_2$H$_4$A Si$_4$H$_8$A $\leftrightarrow$ Si$_2$H$_4$B + Si$_2$H$_4$A	Si$_n$H$_{2m}$A $\leftrightarrow$ Si$_l$H$_{2k}$B + Si$_{n-l}$H$_{2(m-k)}$A	2 $\times$ 10^{15}	$\Delta H_{\text{rxn},1000}$
Silylene to silene isomerization	Si$_2$H$_4$B $\leftrightarrow$ Si$_2$H$_4$A Si$_3$H$_6$B $\leftrightarrow$ Si$_3$H$_6$A	Si$_n$H$_{2m}$B $\leftrightarrow$ Si$_n$H$_{2m}$A	1 $\times$ 10^{13}	7.5 kcal/mol
Ring opening	Si$_3$H$_6$$\leftrightarrow$ Si$_3$H$_6$B Si$_4$H$_8$ $\leftrightarrow$ Si$_4$H$_8$B	Si$_n$H$_{2m}$ $\leftrightarrow$ Si$_n$H$_{2m}$B	2 $\times$ 10^{15}	$\Delta H_{\text{rxn},1000}$

Adapted with permission from Swihart and Girshick (1999). Copyright 1999 American Chemical Society.

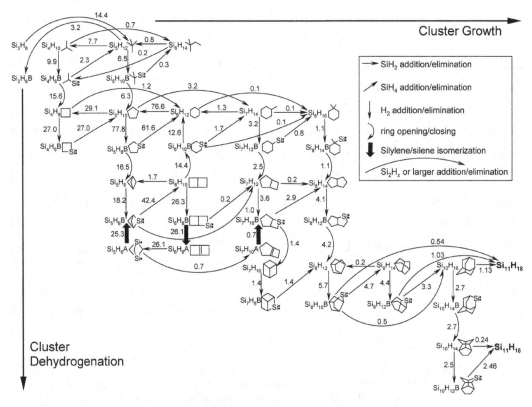

Figure 8.6 Dominant clustering pathway for an initial composition of 1% silane in H$_2$ at 1,023 K, 1 atm, at a reaction time of 5 s, from the numerical simulation by Swihart and Girshick (1999). The form of arrows indicates the type of reaction according to legend, and the numbers next to the arrows indicate the net reaction rate in units of 10^{-15} mol·cm^{-3}·s^{-1}. Reprinted with permission from Swihart and Girshick (1999). Copyright 1999 American Chemical Society.

considered expands dramatically, with associated large increases in computational expense. Moreover, the errors in estimating thermochemical properties are expected to increase as cluster size increases. With the largest cluster size considered in the thermo-chemical database set to 10, reactions that produce silicon hydrides containing more than 10 silicon atoms were assumed to be irreversible. The nucleation rate was thus taken to equal the sum of the production rates of species containing more than 10 Si atoms.

Swihart and Girshick (1999) used their kinetic mechanism to conduct numerical simulations for given conditions of temperature, pressure, and initial mixture compos-ition, for silane diluted in either hydrogen or helium. These simulations allow one to observe the dominant clustering pathways leading to nucleation, as well as the effect of conditions in determining which clustering paths dominate. For example, Figure 8.6 shows the dominant clustering pathways for the case of an initial composition of 1% silane in H_2 at a temperature of 1,023 K and a total pressure of 1 atm, at a time 5 s after the initial introduction of silane, while Figure 8.7 shows the corresponding result with

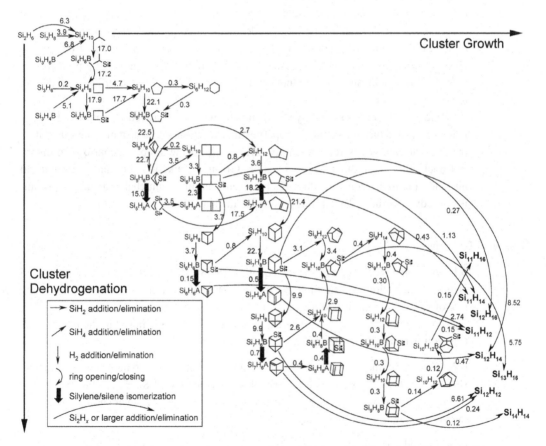

Figure 8.7 Dominant clustering pathway for an initial composition of 1% silane in helium at 1,023 K, 1 atm, at a reaction time of 5 s, from numerical simulation by Swihart and Girshick (1999). The form of arrows indicates the type of reaction according to the legend, and the numbers next to the arrows indicate the net reaction rate in units of 10^{-15} mol·cm^{-3}·s^{-1}. Reprinted with permission from Swihart and Girshick (1999). Copyright 1999 American Chemical Society.

silane diluted in helium rather than hydrogen. In the case of hydrogen dilution, the calculated nucleation rate at 5 s is predicted to equal 3.4×10^9 cm^{-3} s^{-1}, with 96% of that total occurring through the reactions shown in Figure 8.6. In the case of helium dilution, the calculated nucleation rate at 5 s equals 1.6×10^{13} cm^{-3} s^{-1}, with 98% of that total occurring through the reactions shown in Figure 8.7.

These results make it obvious that at least in this case the path to nucleation does not follow the simple repetitive sequence discussed in Section 8.2. It is also interesting to observe, particularly in Figure 8.6, that many of the reactions included in the dominant reaction path have positive net *backward* rates, which remove Si atoms from clusters. Moreover, the dominant reactions in the case of helium dilution are seen to be different than in the hydrogen case. In fact, one finds not only that the nucleation rate itself but even the reactions that dominate the path to nucleation can differ depending on conditions such as temperature and pressure, making it difficult to know which reactions must be included in such a model a priori.

The fact that the nucleation rate is much higher (by a factor of almost 500) in the case of helium dilution than in the case of hydrogen dilution is consistent with experimental observations and is straightforward to understand. Starting with SiH$_4$ decomposition to produce SiH$_2$ and H$_2$, reactions involving hydrogen elimination are a key part of the cluster growth mechanism. Increasing the H:Si ratio in the gas mixture therefore increases the backward rates of these reactions, with the ultimate effect of reducing the nucleation rate.

While the accuracy of these types of models will be improved by improved knowledge of cluster thermochemical properties and reaction rate constants, the qualitative picture that emerges is fundamentally different from the simple scenario of ordinary self-nucleation of a supersaturated vapor. Given the likely importance of chemical nucleation in many contexts, this area represents an important challenge and opportunity for advances in understanding gas-phase nucleation.

Homework Problems

8.1 As noted in the discussion of equation (8.38), the factor α/α_{eq} plays a role in chemical nucleation by a simple repetitive sequence of reactions that is analogous to the role of saturation ratio in ordinary self-nucleation. However, in equation (8.38) α/α_{eq} is raised to the power $-(M-1)$, whereas in equation (2.66) S is raised to the power $-M$. Can you explain where this difference arises from?

8.2 Guided by the example shown in Figure 8.4, use the group additivity parameters shown in Table 8.1 to estimate the enthalpy of formation of Si$_3$H$_6$ at 298.15 K. Note that the structure of Si$_3$H$_6$ is indicated in Figure 8.5, and your answer should agree with the value shown there.

8.3 Consider the reaction

$$Si_3H_8 + Si_3H_6\,B \rightleftarrows Si_6H_{14}.$$

Based on the information in Table 8.2 and Figure 8.5, calculate the value of the forward rate constant $(\text{cm}^3 \cdot \text{s}^{-1})$ of this reaction at 298.15 K.

9   Nucleation in Plasmas

Particle nucleation in plasmas is important in both nature and industry. Much of the universe exists in the plasma state, that is, in the state of an ionized or partially ionized gas, and under appropriate conditions gas-phase nucleation occurs, leading to the formation of interstellar dust and ultimately to the formation of macroscopic bodies. Dust formation in nuclear fusion Tokamak reactors, which may be caused by vapor generation due to the erosion of sections of the reactor walls adjacent to the fusion edge plasma, is one of the critical challenges confronting efforts to develop this technology for future global energy needs. Plasmas are also widely used for many industrial applications, including the manufacture of microelectronics, where unwanted gas-phase nucleation can generate particles that are potential contaminants if they deposit on a semiconductor wafer during processing. Alternatively, plasmas are used for the controlled synthesis of nanoparticles of many substances for a wide variety of applications, ranging from solar cells to cancer treatment.

9.1   Thermal and Nonthermal Plasmas

The different types of plasmas can be broadly classified as being either thermal or nonthermal, according to whether the free electrons in the plasma are in thermal equilibrium with the heavy species (all chemical species other than free electrons). Whether the plasma is thermal or nonthermal can strongly affect the route to particle nucleation.

In thermal plasmas, ionization is created by heating the gas to a high temperature. These temperatures, typically in excess of 6,000 K or even above 10,000 K, are usually sufficient to completely dissociate any molecular species to their elemental constituents. As the gas is transported by convection, and/or species diffuse to colder regions, one or more of these elemental species may become supersaturated or may recombine to form high-temperature molecular species that then become supersaturated. In these colder regions, the degree of ionization is much lower, and the electron temperature, being roughly equilibrated with the gas (i.e., heavy species) temperature, is also lower, compared to conditions in the high-temperature zone. Under such conditions, the supersaturated vapors that form may undergo ordinary self-nucleation, or possibly ion-induced nucleation if the ion concentration is high enough. Thus the "chemical nucleation" scenario discussed in Chapter 8 does not necessarily apply to nucleation in thermal plasma systems.

The situation is different in nonthermal plasmas. Free electrons are accelerated to high velocities by electric fields, and ionization is created mainly by collisions between these free electrons and heavy species. The electrons, being much lighter than the heavy species, may be considerably out of thermal equilibrium with the heavy species, because energy transfer in elastic collisions between bodies of widely disparate mass is much less efficient than between bodies of approximately the same mass. Thus in nonthermal plasmas the electron temperature is typically much higher than the gas temperature. For example, in many types of nonthermal plasmas the gas temperature is around or only somewhat elevated above room temperature, whereas the electron temperature is typically around or greater than one electron volt, where 1 eV corresponds to approximately 11,600 K.

Under these conditions, when the plasma chemical composition is prone to particle formation, it is likely to occur by means of chemical nucleation, particularly if the feedstock reactants are in molecular rather than pure elemental form. However, several basic features are notably different than the scenario discussed in Chapter 8, which considers only nucleation in neutral gases.

This chapter focuses on nucleation in nonthermal plasmas. While nucleation of many substances in plasmas has been studied, as an example we focus on one of the most studied cases, nucleation of silicon particles in silane-containing plasmas. The choice of this system as an example also facilitates comparison with the case of nucleation during the thermal decomposition of silane, discussed in Chapter 8.

9.2   General Features of Nonthermal Plasmas Pertinent to Nucleation

While a detailed discussion of nonthermal plasmas is beyond the scope of this text, we here note several features that are pertinent to nucleation.

Plasmas are generated by applying a high enough electric field to a gas to cause electrical breakdown, thereby causing electrical current to pass through the gas, which in turn facilitates further ionization. In industrial and laboratory plasmas, the plasma is bounded by walls. Free electrons have much greater mobility than heavy species in the plasma, both because they are much lighter and because they are typically at much higher temperatures. Therefore, when a plasma is initiated, electrons diffuse to bounding surfaces much more rapidly than ions, causing these surfaces to become charged to negative potential relative to the bulk plasma. This in turn retards the electron current to the surfaces and increases the current of positive ions, establishing an ambipolar electric field with a positive net space charge. This field exists over a region adjacent to the surface known as an electrical sheath. Within the bulk plasma, overall charge neutrality exists, except within distances smaller than a characteristic length scale known as the Debye length, which is typically much smaller than the dimensions of the plasma itself. Hence this condition is commonly referred to as "quasi-neutrality" of the bulk plasma.

If condensed-phase particles exist in a plasma, they are in effect bounding surfaces. Thus, particles in nonthermal plasmas tend to be negatively charged.[1] However, as they

---

[1] There are situations where particles can be predominantly positively charged, particularly if processes such as thermionic emission or UV-induced photoemission are important.

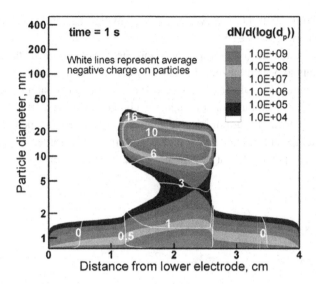

Figure 9.1 Particle size distribution function, 1.0 s after plasma initiation, across a 4cm electrode gap in a parallel-plate radio frequency argon–silane plasma at a pressure of 17 Pa, calculated in a 1-D numerical simulation (Agarwal and Girshick 2012). The asymmetry in the spatial distribution of particles is due primarily to neutral gas drag, with the gas flowing from the upper (showerhead) electrode toward the lower electrode. Reprinted from Agarwal and Girshick (2012) with the permission of IOP Publishing; permission conveyed through Copyright Clearance Center, Inc.

are much more massive than electrons, these negatively charged particles are much less mobile. Thus, being electrostatically repelled from the walls, they are confined by the electric field to the bulk plasma. As particles nucleate and grow, they attach electrons and are pulled by the electric field toward the center of the plasma. The larger the particle, the more electrons it can hold. At very small sizes, at most a few nanometers in diameter (depending on conditions and the particle material), electron tunneling limits the number of electrons that a particle can hold to at most one or two. As these nanoparticles collide with and attach electrons and ions, they experience a fluctuating charge such that at any time a fraction of the particles are neutral or even positively charged. These nonnegative particles are no longer confined by the electric field to the bulk plasma and can diffuse to the walls. This diffusional loss can be important, particularly in the low-pressure environments that are typical of nonthermal plasmas used in industrial applications such as semiconductor processing.

Figure 9.1, from a one-dimensional numerical simulation of a parallel-plate radio frequency argon–silane plasma at a pressure of 17 Pa, illustrates these qualitative features (Agarwal and Girshick 2012). This figure shows the profile of the particle size distribution across a 4cm electrode gap, 1.0 s after initiation of the plasma. Shaded contours show the particle size distribution function $dN/d\log(D_p)$, where N is the total number of particles and D_p is the particle diameter, while white lines show the particles' average negative charge. At the midplane of the gap, the peaks in the size distribution function occur at diameters of ~15 nm and smaller than 1 nm. One sees that particles having diameters of ~10–20 nm have an average negative charge also of the order ~10 and are confined to the bulk plasma, while particles smaller than about 2 nm, with an average charge close to zero, are spread more uniformly across the gap, due to diffusion.

As particles grow and become on average more negatively charged, they become increasingly unlikely to be neutralized, and thus they become stably confined to the bulk plasma and concentrated in the region where the electric potential profile has its maximum value. The accumulation of negative charge on these particles depletes electrons from the plasma. In turn, this can change plasma properties such as electron density, electron temperature, ion density, and electric potential profile.

The scalar electric potential ϕ at any point in the plasma is related to the vector electric field $\mathbf{E}$ by

$$\mathbf{E} = -\nabla\phi, \tag{9.1}$$

and the potential is itself governed by Poisson's equation,

$$\nabla^2\phi = -\frac{\rho^c}{\varepsilon_0}, \tag{9.2}$$

where ρ^c is the net charge density. In turn, the net charge density equals the sum of the charge densities (which may be positive or negative) of all charge carriers, including electrons, ions, and condensed-phase particles. Assuming that all ions are singly ionized,[2] this can be written as

$$\rho^c = e\left(N_+ - N_e - N_- - N_p z_{p(-)}\right), \tag{9.3}$$

where e is the elementary charge, $N_+, N_e, N_-,$ and N_p are, respectively, the number densities of positive ions, electrons, negative ions, and particles, and $z_{p(-)}$ is the average particle negative charge.

Thus the existence of condensed-phase particles affects the plasma properties, which in turn changes the environment that governs particle nucleation, growth, and transport. For example, electron temperature and electron density are the key drivers of chemical reactions that govern particle nucleation. The plasma and the aerosol phase in such systems can be strongly coupled.

9.3   Nucleation of Silicon Particles in Silane-Containing Plasmas

9.3.1   Plasma-Driven vs. Thermally Driven Nucleation in Silane

Much of the discussion in Section 8.3.2 about particle nucleation during thermal decomposition of silane is pertinent to the situation in nonthermal plasmas. However, there are some important differences.

First, thermal decomposition decomposes SiH_4 primarily into SiH_2 and H_2, and the relative abundances of monatomic H, SiH_3, and higher silyl species at temperatures that are typically employed are negligibly small. Effectively, as clusters grow, only Si_nH_m species for which m is an even number exist.

---

[2] The extension to account for multiply charged ions is straightforward.

In contrast, in nonthermal plasmas silane is dissociated mainly by electron impact, because under typical conditions, the gas temperature is too low to cause significant dissociation, while the electron temperature is much higher. Electron-impact dissociation of silane occurs by both

$$SiH_4 + e \rightarrow SiH_3 + H + e \qquad (9.4)$$

and

$$SiH_4 + e \rightarrow SiH_2 + 2H + e. \qquad (9.5)$$

Both of these reactions generate monatomic H, which can also be generated by electron-impact dissociation of H_2. Thus monatomic H, which exists in negligible abundance in the thermal decomposition of silane, is an important species in plasma decomposition. In turn, monatomic H can abstract H atoms from silane,

$$SiH_4 + H \rightarrow SiH_3 + H_2, \qquad (9.6)$$

as well as from higher silanes, leading to significant abundances of highly reactive silyl radicals, Si_nH_m with m odd-numbered.

These species undergo further reactions, generating a mixture that includes silicon hydrides of various stoichiometries Si_nH_m, in both neutral and ionic forms. Experimental studies by Howling, Hollenstein, and coworkers demonstrated that silicon hydride anion clusters can grow to large sizes in RF silane plasmas, while corresponding neutral and cation clusters are limited to much smaller sizes (Howling et al. 1993a, 1993b, 1994, 1996). An example is shown in Figure 9.2, which shows mass spectrometry measurements of silicon hydride anions and cations extracted from a modulated radio frequency silane plasma at a pressure of 0.1 mbar (Howling et al. 1994). Each peak in these spectra corresponds to an ion with a given number of silicon atoms, while the width of the curve associated with each peak corresponds to a variable number of hydrogen atoms. The data shown in Figure 9.2 are raw intensities, uncorrected for the transmission efficiency of the instrument, which decreases with increasing cluster mass. However, as the transmission efficiency is the same for ions of either polarity, it can be seen that negative ions exist in the plasma up to much larger sizes than positive ions.

These results are consistent with the picture presented in Section 9.2. Negative silicon hydride ions are confined to the bulk plasma by the electric field, allowing them time to grow, whereas their positive counterparts are expelled by the ambipolar field and lost before they can grow. Meanwhile, neutral silicon hydrides, being unaffected by the electric field, can be lost by diffusion under the low-pressure conditions typically employed, and thus they too are unlikely to grow to large sizes. Even if some neutral clusters do grow to larger sizes, as they grow they become increasingly likely to attach an electron and join the population of anion clusters.

Growth of silicon hydride anions is expected to occur by reactions with neutral Si-containing species, of which the most likely candidates are SiH_4 and SiH_m radicals, the most prominent of these being SiH_3 and SiH_2. SiH_4 is relatively unreactive but, depending on its degree of dissociation, is often by far the most abundant Si-containing species in the plasma, whereas SiH_m radicals are highly reactive but typically much less abundant.

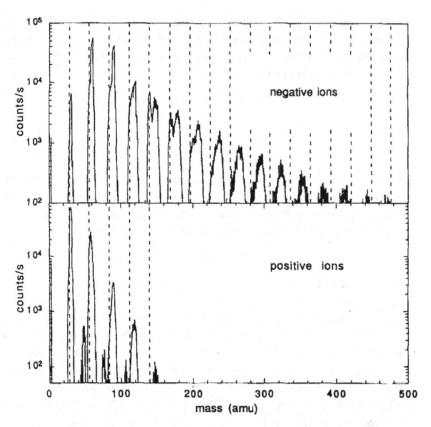

Figure 9.2 Mass spectra intensities of negative and positive silicon hydride ion groups for a modulated radio frequency silane plasma at a pressure of 0.1 mbar (Howling et al. 1994). Each group represents Si_nH_m with a given value of n but variable values of m. Reprinted from Howling et al. (1994) with the permission of AIP Publishing.

In addition to the trapping of anions by the electric field, there are several factors that favor this anion-neutral growth path: Anion–anion reactions are unlikely due to mutual Coulomb repulsion; neutrals are typically far more abundant in these types of plasma than silicon hydride cations; and ion–molecule reactions are generally faster than the corresponding neutral–neutral reactions, due to the attractive effect of induced dipoles.

9.3.2   Thermodynamic Properties of Anions

Under many conditions that occur in nonthermal plasmas, reactions driven by electron impact are often effectively irreversible. However, this does not apply to the reactions involved in the growth of silicon hydride clusters, discussed in Section 9.3.1, for which reversibility is an important feature that can determine under what conditions nucleation does or does not occur to any significant degree. One thus requires information on the equilibrium constants of each of the reactions in the cluster growth mechanism, which in turn requires information on the Gibbs free energies of all species participating in these reactions.

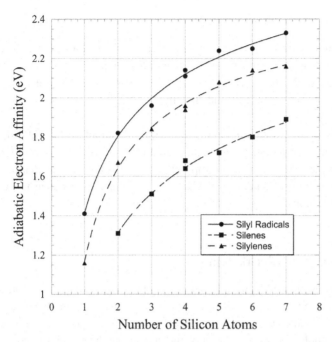

Figure 9.3 Electron affinities in eV of silicon hydrides, calculated by Swihart (2000) using density functional theory. Each data point shows the result for the most stable isomer of the acyclic terminal species of each type. Adapted with permission from Swihart (2000). Copyright 2020 American Chemical Society.

As noted above, most cluster growth reactions in this system are expected to involve silicon hydride anions, $Si_nH_m^-$, reacting with neutral silicon hydrides containing one or two Si atoms. Data are available for the thermodynamic properties of the small neutral silicon hydrides, and also, as discussed in Section 8.3.2, data or at least reasonable estimates are available for the thermodynamic properties of neutral Si_nH_m, for n up to about 10.

The difference between the energy (and hence the enthalpy, at given temperature) of an anion and its parent neutral is termed the electron affinity. As it requires work to remove an electron from an anion, if the electron affinity is positive then the enthalpy of the anion is lower by that amount than the enthalpy of the parent neutral.

Swihart (2000) used density functional theory to calculate the electron affinities of 72 unsaturated silicon hydrides Si_nH_m, for $n = 1$ to 7 and $m = 0$ to 15. The species examined were the silyl radicals, silylenes, and silene species discussed in Section 8.3.2 that are believed to be important in cluster growth. Saturated silicon hydrides such as SiH_4, Si_2H_6, etc., were not included in the calculations because they have negative electron affinities, meaning that they do not form stable anions. Swihart estimated that his calculated electron affinities are accurate to within 0.15 eV. Figure 9.3 shows results for the calculated electron affinities in eV for acyclic terminal silyl radicals, silenes, and silylenes. As can be seen, for each type of species the electron affinity systematically increases with cluster size. Moreover, the data for all three types of species are seen to be well fitted by curves of the form

$$EA = a - bn^c, \qquad (9.7)$$

where EA is the electron affinity, n is the number of silicon atoms in the cluster, and a, b, and c are constants that produce the best curve fits for each species type. For the silyl radicals, $a = 3.42$ eV, $b = 2.00$ eV, and $c = -0.314$; for the silenes, $a = 3.36$ eV, $b = 2.45$ eV, and $c = -0.256$; and for the silylenes, $a = 2.74$ eV, $b = 1.58$ eV, and $c = -0.518$.

These curves were extrapolated to size $n = 10$ by Bhandarkar et al. (2000), who then used these results to calculate the enthalpies of all silicon hydride anions in their kinetic mechanism. By assuming that the entropy of the anion is the same as that of the parent neutral, these electron affinity values then allowed them to calculate the Gibbs free energies of the anions.

9.3.3   Kinetics of Anion Cluster Growth

As an ion and a neutral molecule approach each other, attractive polarization forces can bring the neutral molecule into orbit around the ion, which may finally result in contact. The resulting collision rate is given by Langevin's theory. For the case where the neutral molecule is nonpolar (including SiH_4 and higher silanes), the rate constant for collisions is given by

$$k_L = e \sqrt{\frac{\pi \alpha}{\varepsilon_0 m_r}}, \qquad (9.8)$$

where k_L is the Langevin rate constant in MKSA units, α is the polarizability of the neutral molecule in Å^3, and m_r is the reduced mass of the ion-neutral collision pair in atomic mass units. For SiH_4, which is typically the most abundant Si-containing neutral in nonthermal silane plasmas, the polarizability equals 4.62 Å^3, and for Si_2H_6, $\alpha = 8.47$ Å^3. Perrin et al. (1996) suggested a scaling law for extrapolating that value to neutral silicon hydrides with stoichiometry Si_nH_{2n+1},

$$\alpha(Si_n H_{2n+1}) = \alpha(SiH_4)[1 + 0.8(n - 1)]. \qquad (9.9)$$

For polar molecules, including the radicals SiH, SiH_2, and SiH_3, k_L includes an additional term proportional to the permanent dipole moment of the molecule. Perrin et al. (1996) note that this term adds at most 20% to the value of k_L for gas temperatures in the range 300–500 K.

The Langevin rate constant for collisions of SiH_m species with SiH_3^- equals approximately $10^{-9} \text{cm}^3 \cdot \text{s}^{-1}$ (Watanabe et al. 1996), and the actual reaction rate constant with radical species SiH_3 or SiH_2 as collision partner is believed to be close to this value, meaning that the reaction is barrierless and almost all collisions result in reaction (Gallagher et al. 2002). However, for SiH_4 as a collision partner the actual reaction rate constant is much lower, estimated as $4 - 10 \times 10^{-12} \text{cm}^3 \cdot \text{s}^{-1}$ (Howling et al. 1994; Perrin et al. 1996). Thus SiH_4 is less reactive for anion cluster growth than radical SiH_m species by two or three orders of magnitude.

Bao et al. (2015) used computational chemistry[3] to estimate rate constants for the initial reactions lying at the root of the two clustering sequences predicted by Bhandarkar et al. (2000) to dominate nucleation:

$$Si_2H_4^- + SiH_4 \rightarrow Si_3H_6^- + H_2 \tag{9.10}$$

and

$$Si_2H_5^- + SiH_4 \rightarrow Si_3H_7^- + H_2. \tag{9.11}$$

Bao et al. proposed that each of these reactions follows a three-step mechanism, with each of the steps having significant activation energies and thermal rate constants many orders of magnitude lower than the upper bound given by Langevin's theory.

Under many conditions, SiH_4 may be much more abundant than SiH_3 or SiH_2. For example, in three different studies that reported calculated species number densities based on chemical kinetic modeling of silane-containing low-pressure RF plasmas, the SiH_4-to-SiH_3 abundance ratio ranged from ~10^2 (Gallagher et al. 2002), to ~10^3 (De Bleecker et al. 2004), to ~10^4 (Agarwal and Girshick 2014). While there are some differences in the kinetic mechanisms employed in these studies, the range of results for the SiH_4-to-SiH_3 abundance ratio can be ascribed mainly to the different plasma conditions considered, such as inlet gas (pure silane or silane diluted in noble gas), applied RF voltage, interelectrode spacing, and pressure. Thus it is not remarkable that these authors found quite different results for the calculated SiH_4-to-SiH_3 abundance ratio. With the caveat that there appears to be considerable uncertainty in the rate constant for SiH_4 as a growth species, one can therefore conclude that both SiH_4 and radical species, including SiH_3 and, to a lesser extent, SiH_2, may contribute significantly to cluster growth, with the relative contributions depending on plasma conditions.

Moreover, the conclusion that SiH_4 is much less reactive as an anion cluster growth species than SiH_3 (Perrin et al. 1996) is based on the assumption that SiH_4 is in vibrational equilibrium at the typical gas temperature of ~300–500 K, for which SiH_4 would be overwhelmingly in its vibrational ground state. However, in plasmas electron impact can drive vibrational nonequilibrium, and vibrationally excited SiH_4 would be expected to be much more reactive than ground-state SiH_4. Indeed Fridman et al. (1996) proposed that the clustering sequence given by reactions (8.1) to (8.4) was driven primarily by vibrationally excited SiH_4.

The relative contributions to cluster growth of SiH_3, SiH_4, and vibrationally excited SiH_4 were estimated in the chemical kinetic model of De Bleecker et al. (2004), as shown in Figure 9.4. Noting the considerable uncertainties in the rate constant for anion clustering with ground-state SiH_4, they varied its rate constant from 10^{-10} to 10^{-12} cm^3 s^{-1}, roughly corresponding to the range spanning from 10 to 1000 times lower than the Langevin collision rate constant. For their assumed conditions, they found that at the lower end of the range the relative contributions to cluster growth were fairly evenly divided between ground-state SiH_4 (38%), vibrationally excited SiH_4

---

[3] Specifically, they used multistructural canonical variational transition state theory with the small-curvature tunneling approximation. See Bao et al. (2015).

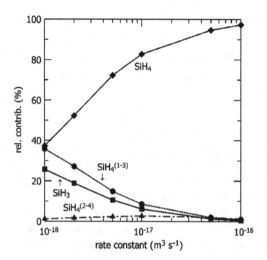

Figure 9.4 Relative contributions of various neutral species to anion cluster growth in a silane plasma, as a function of the assumed rate constant for cluster growth with ground-state SiH_4 as the neutral reactant, calculated by De Bleecker et al. (2004). $SiH_4(2–4)$ and $SiH_4(1–3)$ refer, respectively, to the bending and stretching modes of silane in its first vibrational level. Reprinted with permission from De Bleecker et al. (2004). Copyright 2004 by the American Physical Society.

(36%), and SiH_3 (25%), while at the upper end ground-state SiH_4 provided nearly 100% of growth.

Detailed numerical simulations by Bhandarkar et al. (2000) suggest that the dominant pathways to anion cluster growth are more complicated than the simple scenario of a single repetitive sequence, at least for conditions where the SiH_m radical abundance is ~10^4 lower than that of SiH_4. In that work, the thermochemical properties and chemical kinetic mechanism for the case of thermal decomposition of silane discussed in Section 8.3.2 were extended to include reactions and species important in plasma decomposition, including electrons and ions, monatomic H, and silicon hydrides containing odd numbers of hydrogen atoms.

For a gas temperature of 500 K and a positive ion density of $3 \times 10^9 \, \text{cm}^{-3}$, Figure 9.5 shows the dominant types of cluster growth reactions predicted by this model, while Figures 9.6 and 9.7 show the dominant anionic reaction pathways and neutral reaction pathways, respectively. The number shown for each reaction is the net reaction rate at steady state, in units of $10^{-15} \, \text{mol} \cdot \text{cm}^{-3} \cdot \text{s}^{-1}$.

As can be seen, the net reaction rates for the growth of anion clusters were generally predicted to be higher by several orders of magnitude compared to the growth of neutral clusters, consistent with the experimental observations and qualitative features discussed above. In these and other numerical studies, the nucleation rate was set equal to the net production rate of clusters containing an arbitrary number of Si atoms, for example, greater than 10 Si atoms (Agarwal and Girshick 2014; Bhandarkar et al. 2000, 2003) or greater than 12 Si atoms (De Bleecker et al. 2004a, 2004b). In Figure 9.6, it can be seen that the dominant nucleation paths for the conditions considered involve clustering chains that start with anionic silylene ($Si_2H_4^-$) or silyl ($Si_2H_5^-$). Based on

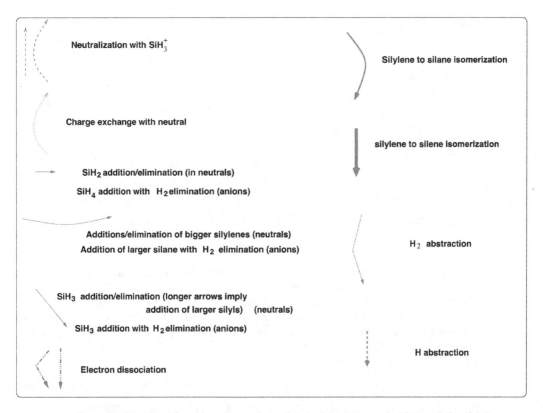

Figure 9.5 Meaning of various arrows shown in reaction pathways in Figures 9.6 and 9.7. Reprinted from Bhandarkar et al. (2000) with the permission of IOP Publishing; permission conveyed through Copyright Clearance Center, Inc.

those results, in two-dimensional simulations in which the number of species that could be considered was constrained by computational expense, Le Picard et al. (2016) equated the nucleation rate to the sum of the rates of reactions in which one Si atom was added to either $Si_2H_4^-$ or $Si_2H_5^-$.

As in the cases of chemical nucleation in thermally driven systems discussed in Section 8.3, the choice in all of these studies of the minimum cluster size to treat as a "particle" was arbitrary and does not imply the existence of a critical cluster size. The choice of the maximum cluster size to treat in a chemical kinetic nucleation scheme reflects a trade-off between the conflicting demands of accuracy and computational expense. Aside from the computational expense, increasing the minimum cluster size to treat as a "particle" does not necessarily imply increased accuracy, as uncertainties in the thermodynamic properties and reactivities of clusters may increase as cluster size increases.

9.3.4   Temperature Dependence of Time for the Onset of Nucleation

Boufendi and Bouchoule (1994) made an interesting experimental observation on particle formation in nonthermal silane-containing plasmas. Figure 9.8 shows their

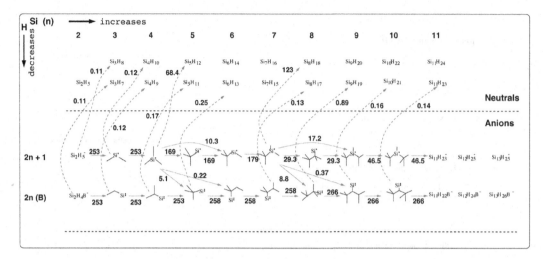

Figure 9.6 Dominant anionic reaction pathways at steady state predicted by a 0-D numerical model for a capacitively coupled radio frequency silane plasma. Conditions assumed: pressure $= 100$ mTorr, frequency $= 13.56$ MHz, gas temperature $= 500$ K, and positive ion density $= 3 \times 10^9$ cm^{-3}. Numbers indicate net reaction rates in units of 10^{-15} mol $\cdot$ cm$^{-3} \cdot$ s^{-1}. Reprinted from Bhandarkar et al. (2000) with the permission of IOP Publishing; permission conveyed through Copyright Clearance Center, Inc.

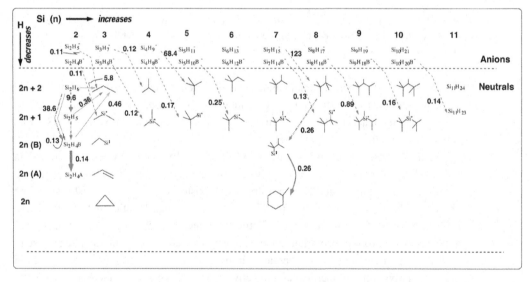

Figure 9.7 Dominant neutral reaction pathways at steady state predicted by a 0-D numerical model for a capacitively coupled radio frequency silane plasma. Same conditions as in Figure 9.6. Numbers indicate net reaction rates in units of 10^{-15} mol $\cdot$ cm$^{-3} \cdot$ s^{-1}. Reprinted from Bhandarkar et al. (2000) with the permission of IOP Publishing; permission conveyed through Copyright Clearance Center, Inc.

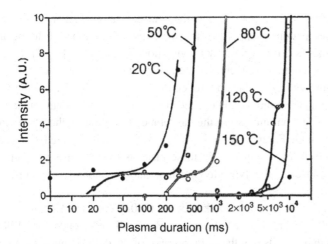

Figure 9.8 Signal intensity from laser-induced particle explosive evaporation, with various gas temperatures, in the experiments of Boufendi and Bouchoule (1994). The signal is proportional to the product of particle number density and the fourth power of particle diameter. Reprinted from Boufendi and Bouchoule (1994) with the permission of IOP Publishing; permission conveyed through Copyright Clearance Center, Inc.

experimental results for the signal intensity from laser-induced particle explosive evaporation (LIPEE), versus plasma duration time, for gas (heavy species) temperatures ranging from 20 to 150°C. The LIPEE signal intensity I_{LIPEE} varies as

$$I_{LIPEE} \propto N_p d_p^4, \tag{9.12}$$

N_p and d_p being, respectively, the particle number density and diameter.

As can be seen in Figure 9.8, at each temperature the LIPEE signal exhibited a sudden steep rise at some time, which can be interpreted as the time required for the onset of observable particle formation. Increasing the gas temperature from 20 to 150°C increased the nucleation onset time by approximately two orders of magnitude.

In ordinary self-nucleation from a supersaturated vapor, increasing the temperature reduces the vapor saturation ratio, assuming a fixed partial pressure of the condensible vapor, and therefore reduces the nucleation rate, which seems qualitatively consistent with the trend in Figure 9.8, insofar as the time required for the onset of observable nucleation is negatively correlated with the nucleation rate.

However, as discussed above, nucleation in silane plasmas involves chemical nucleation not ordinary self-nucleation from a supersaturated vapor. Moreover, depending on conditions such as dilution gas and pressure, thermal dissociation of silane typically requires temperatures above about 700°C, and increasing the gas temperature above that increases the degree of silane dissociation and hence *increases* the nucleation rate.

In contrast, silane in nonthermal plasmas is dissociated by electron impact not thermally, so it is not immediately obvious why increasing the heavy species temperature should so strongly affect the onset time for nucleation.

Fridman et al. (1996) proposed that the increase in nucleation onset time with increasing gas temperature supported their hypothesis that anion clustering in silane-containing plasmas is driven primarily by vibrationally excited SiH_4. They argued that increasing the gas temperature increases the rate of vibrational relaxation, as described by Landau–Teller theory (Landau and Teller 1936), and thereby reduces the fraction of SiH_4 molecules that are vibrationally excited. Therefore, insofar as vibrationally excited silane induces nucleation, increasing the gas temperature reduces the nucleation rate.

On the other hand, Figure 9.4 suggests that vibrationally excited silane does not contribute enough to cluster growth in these plasmas to explain this strong effect of temperature, because of the typically much greater abundance of neutral ground-state SiH_4.

To address this problem, Bhandarkar et al. (2003) conducted detailed numerical simulations of the experiments conducted by Boufendi and Bouchoule (1994). They concluded from their results that the primary cause of the delay in nucleation onset time with increasing gas temperature was the effect of temperature on Brownian diffusion.

Hard-sphere kinetic theory of gases predicts that for a single-component gas the coefficient of diffusion (thus, the coefficient of self-diffusion) varies, to first approximation, as (Vincenti and Kruger 1965)

$$D \propto \frac{T^{1/2}}{N}. \tag{9.13}$$

For an ideal gas at fixed pressure, number density is inversely proportional to temperature. Therefore, at fixed pressure,

$$D \propto T^{3/2}. \tag{9.14}$$

While the temperature dependence of the diffusion coefficient for a species in a multicomponent mixture is more complicated than this, equation (9.14) is still qualitatively correct.

As discussed in Section 9.3.1, cluster growth in nonthermal silane-containing plasmas primarily involves the growth of anions by reactions with neutral Si-containing species, where the anion clusters are trapped by the electric field in the bulk plasma while the neutrals are free to diffuse. Therefore, at higher gas temperatures, the more-than-linear effect of gas temperature on the diffusivity of neutral species causes their concentration in the bulk plasma to be significantly reduced, slowing the cluster growth rate and thus delaying the onset of observable particle formation. This applies to all neutral species not just clusters. The effect is especially noteworthy for monatomic hydrogen. Being the lightest of the heavy species, monatomic H is the most diffusive and is also an important driver of much of the chemistry responsible for cluster growth, for example, via reactions such as reaction (9.6).

9.4   Nucleation in Plasmas: Conclusions

Gas-phase nucleation in plasmas occurs in many different contexts, ranging from interstellar space and nuclear fusion reactors to semiconductor processing and nanoparticle synthesis. In some cases, such as thermal plasma synthesis of metal nanoparticles, nucleation may follow the conventional scenarios of single-component homogeneous or ion-induced nucleation. In other cases, such as dust formation in nonthermal plasmas of the type used in semiconductor processing, the paths to nucleation are specific to the chemistry of the gases introduced into the processing chamber. In such cases, nucleation is likely to involve a detailed mix of phenomena that combine chemical nucleation, discussed in Chapter 8, with plasma physics, particularly with regard to the charging of small clusters that constitute the path to particle formation. These charged clusters consume electrons and interact with the electric field in the plasma, leading to a coupling between the plasma and the aerosol phase.

The example presented in this chapter, of silicon particle nucleation in silane-containing plasmas, has been the most studied because of its importance in semiconductor processing. However, other chemical systems used in the semiconductor industry, such as fluorocarbon plasmas used for plasma etching, are also known to be prone to powder formation. Developing accurate predictive models of gas-phase nucleation can only be accomplished by developing a detailed understanding of the thermochemistry and kinetics of cluster growth in the specific chemical systems involved, together with their interaction with the specific plasma environment.

As with the chemical nucleation of soot in hydrocarbon combustion, many of the phenomena involved in nucleation in plasmas may seem far removed from the scenario of single-component homogeneous nucleation that has occupied most of the gas-phase nucleation literature. Yet gas-phase nucleation certainly does occur in fires, internal combustion engines, semiconductor processing plasmas, and many other important contexts where cluster growth is governed by chemical reactions and/or plasma physics. Thus any comprehensive view of the subject of gas-phase nucleation must embrace such seemingly disparate regimes.

Ultimately, one can anticipate that a more unified view of gas-phase nucleation in these quite different environments – from single-component self-nucleation to chemical nucleation in plasmas – will be achieved by utilizing advances in atomistic approaches to model the "conventional" scenarios that to date have mainly been viewed through the lens of classical nucleation theory.

Homework Problems

9.1 Consider Figure 9.1, which shows the results of a numerical simulation of the spatial profiles of particle size distributions and average negative charge across the electrode gap in an argon–silane plasma from Agarwal and Girshick (2012). In the center of the plasma, 2 cm above the lower electrode, it appears that the number density of 5-nm-diameter particles is considerably smaller than that of either much smaller (e.g., 1 nm) or much larger (e.g., 10–20 nm) particles. Can you give a physical explanation for this behavior?

9.2   Consider the silylene anion $Si_3H_6B^-$. Based on information in Figures 8.5 and 9.3, what is the value of its standard Gibbs free energy of formation, $\Delta_f G^0$, at 298.15 K?

9.3   Based on the scaling law given by equation (9.9), what is the value of the Langevin rate constant for the reaction

$$SiH_3^- + Si_3H_8 \rightarrow Si_4H_9^- + H_2?$$

References

Abraham FF (1969). Multistate kinetics in nonsteady-state nucleation: A numerical solution. *Journal of Chemical Physics* **51**, 1632–1638. doi:10.1063/1.1672224.

Abraham FF (1974). *Homogeneous Nucleation Theory: The Pretransition Theory of Vapor Condensation*. New York and London: Academic Press.

Adams GW, Schmitt JL, and Zalabsky RA (1984). The homogeneous nucleation of nonane. *Journal of Chemical Physics* **81**, 5074–5078. doi:10.1063/1.447496.

Afzalifar A, Shields GC, Fowler VR, and Ras RHA (2022). Probing the free energy of small water clusters: Revisiting classical nucleation theory. *Journal of Chemical Physics Letters* **13**, 8038–8046. doi:10.1021/acs.jpclett.2c01361.

Agarwal P and Girshick SL (2012). Sectional modeling of nanoparticle size and charge distributions in dusty plasmas. *Plasma Sources Science and Technology* **21**, 055023. doi:10.1088/0963-0252/21/5/055023.

Agarwal P and Girshick SL (2014). Numerical modeling of the spatiotemporal behavior of an RF argon-silane plasma with dust particle nucleation and growth. *Plasma Chemistry and Plasma Processing* **34**, 489–503. doi:10.1007/s11090-013-9511-3.

Alcock CB, Itkin VP, and Horrigan MK (1984). Vapour pressure equations for the metallic elements: 298–2500 K. *Canadian Metallurgical Quarterly* **23**, 309–313. doi:10.1179/cmq.1984.23.3.309.

Andres RP and Boudart M (1965). Time lag in multistate kinetics: Nucleation. *Journal of Chemical Physics* **42**, 2057–2064.

Anisimov MP, Hopke PK, Shandakov SD, and Shvets II (2000). *n*-Pentanol–helium homogeneous nucleation rates. *Journal of Chemical Physics* **113**, 1971–1975. doi:10.1063/1.482002.

Bao JL, Seal P, and Truhlar DG (2015). Nanodusty plasma chemistry: A mechanistic and variational transition state theory study of the initial steps of silyl anion-silane and silylene anion-silane polymerization reactions. *Physical Chemistry Chemical Physics* **17**, 15928–15935. doi:10.1039/c5cp01979f.

Becker R and Döring W (1935). Kinetische Behandlung der Keimbildung in übersättigten Dämpfen [Kinetic treatment of nucleation in supersaturated vapors]. *Annalen der Physik* **24**, 719–752.

Benson SW (1976). *Thermochemical Kinetics: Methods for the Estimation of Thermochemical Data and Rate Parameters*, 2nd ed. New York: Wiley.

Bhabhe A and Wyslouzil B (2011). Nitrogen nucleation in a cryogenic supersonic nozzle. *Journal of Chemical Physics* **135**, 244311. doi:10.1063/1.3671453.

Bhandarkar U, Swihart MT, Girshick SL, and Kortshagen U (2000). Modeling of silicon hydride clustering in a low-pressure silane plasma. *Journal of Physics D: Applied Physics* **33**, 2731–2746. doi:10.1088/0022-3727/33/21/311.

Bhandarkar U, Kortshagen U, and Girshick SL (2003). Numerical study of the effect of gas temperature on the time for onset of particle nucleation in argon-silane low pressure plasmas. *Journal of Physics D: Applied Physics* **36**, 1399–1408. doi:10.1088/0022-3727/36/12/307.

Blander M and Katz JL (1972). The thermodynamics of cluster formation in nucleation theory. *Journal of Statistical Physics* **4**, 55–59.

Blanquart G and Pitsch H (2009). Analyzing the effects of temperature on soot formation with a joint volume-surface-hydrogen model. *Combustion and Flame* **156**, 1614–1626. doi:10.1016/j. combustflame.2009.04.010.

Blanquart G, Pepiot-Desjardins P, and Pitsch H (2009). Chemical mechanism for high temperature combustion of engine relevant fuels with emphasis on soot precursors. *Combustion and Flame* **156**, 588–607. doi:10.1016/j.combustflame.2008.12.007.

Blokhuis EM and Kuipers J (2006). Thermodynamic expressions for the Tolman length. *Journal of Chemical Physics* **124**, 074701. doi:10.1063/1.2167642.

Boufendi L and Bouchoule A (1994). Particle nucleation and growth in a low-pressure argon-silane discharge. *Plasma Sources Science and Technology* **3**, 262–267. doi:10.1088/0963-0252/3/3/004.

Brus D, Hyvarinen AP, Zdimal V, and Lihavainen H (2005). Homogeneous nucleation rate measurements of 1-butanol in helium: A comparative study of a thermal diffusion cloud chamber and a laminar flow diffusion chamber. *Journal of Chemical Physics* **122**, 214506. doi:10.1063/1.1917746.

Brus D, Zdimal V, and Stratmann F (2006). Homogeneous nucleation rate measurements of 1-propanol in helium: The effect of carrier gas pressure. *Journal of Chemical Physics* **124**, 164306. doi:10.1063/1.2185634.

Brus D, Hyvarinen AP, Zdimal V, and Lihavainen H (2008a). Erratum: "Homogeneous nucleation rate measurements of 1-butanol in helium: A comparative study of a thermal diffusion cloud chamber and a laminar flow diffusion chamber" [vol 122, pg 214506, 2005]. *Journal of Chemical Physics* **128**, 079901. doi:10.1063/1.2830800.

Brus D, Zdimal V, and Smolik J (2008b). Homogeneous nucleation rate measurements in supersaturated water vapor. *Journal of Chemical Physics* **129**, 174501. doi:10.1063/1.3000629.

Brus D, Zdimal V, and Smolik J (2008c). Supplemental information to *Journal of Chemical Physics* 129, 174501 (2008): Homogeneous nucleation rate measurements in supersaturated water vapor. EPAPS Document No. E-JCPSA6-129-614841. doi: 10.1063/1.3151622.

Brus D, Zdimal V, and Uchtmann H (2009). Homogeneous nucleation rate measurements in supersaturated water vapor II. *Journal of Chemical Physics* **131**, 074507. doi:10.1063/1.3211105.

Bumstead HA and Van Name RG (eds) (1906). *The Scientific Papers of J. Willard Gibbs*, vol 1. London: Longmans, Green and Co, 252–258.

Campagna MM, Hruby J, van Dongen MEH, and Smeulders DMJ (2020). Homogeneous water nucleation: Experimental study on pressure and carrier gas effects. *Journal of Chemical Physics* **153**, 164303. doi:10.1063/5.0021477.

Chase MW (1998). *NIST-JANAF Thermochemical Tables*. Washington, DC: American Chemical Society; American Institute of Physics for the National Institute of Standards and Technology.

Connelly BC, Long MB, Smooke MD, Hall RJ, and Colket MB (2009). Computational and experimental investigation of the interaction of soot and NO in coflow diffusion flames. *Proceedings of the Combustion Institute* **32**, 777–784. doi:10.1016/j.proci.2008.06.182.

D'Anna A (2009). Combustion-formed nanoparticles. *Proceedings of the Combustion Institute* **32**, 593–613. doi:10.1016/j.proci.2008.09.005.

De Bleecker K, Bogaerts A, Gijbels R, and Goedheer W (2004a). Numerical investigation of particle formation mechanisms in silane discharges. *Physical Review E* **69**, 056409. doi:10.1103/PhysRevE.69.056409.

De Bleecker K, Bogaerts A, and Goedheer W (2004b). Modeling of the formation and transport of nanoparticles in silane plasmas. *Physical Review E* **70**, 056407. doi:10.1103/PhysRevE.70.056407.

Dingilian KK, Lippe M, Kubecka J, Krohn J, Li CX, Halonen R, Keshavarz F, Reischl B, Kurten T, Vehkamäki H, Signorell R, and Wyslouzil BE (2021). New particle formation from the vapor phase: From barrier-controlled nucleation to the collisional limit. *Journal of Physical Chemistry Letters* **12**, 4593–4599. doi:10.1021/acs.jpclett.1c00762.

Dobbins RA, Eklund TI, and Tjoa R (1977). The direct measurement of the nucleation rate constants. In Pouring AA (ed), *Condensation in High Speed Flows*. New York: American Society for Mechanical Engineers, 43–58.

Dobbins RA, Eklund TI, and Tjoa R (1980). Direct measurement of the nucleation rate constants. *Journal of Aerosol Science* **11**, 23–33. doi:10.1016/0021-8502(80)90141-X.

Elm J, Kubecka J, Besel V, Jaaskelainen MJ, Halonen R, Kurten T, and Vehkamäki H (2020). Modeling the formation and growth of atmospheric molecular clusters: A review. *Journal of Aerosol Science* **149**. 105621. doi:10.1016/j.jaerosci.2020.105621.

Farkas L (1927). Keimbildungsgeschwindigkeit in übersättigten Dämpfen [Nucleation rate in supersaturated vapors]. *Zeitschrift für physikalische Chemie, Stöchiometrie und Verwandtschaftslehre* **125 U**, 236–242. doi:10.1515/zpch-1927-12513.

Feder J, Russell KC, Lothe J, and Pound GM (1966). Homogeneous nucleation and growth of droplets in vapours. *Advances in Physics* **15**, 111–178. doi:10.1080/00018736600101264.

Flagan RC (2007). A thermodynamically consistent kinetic framework for binary nucleation. *Journal of Chemical Physics* **127**, 214503. doi:10.1063/1.2800001.

Flageollet-Daniel C, Garnier JP and Mirabel P (1983). Microscopic surface tension and binary nucleation. *Journal of Chemical Physics* **78**, 2600–2606. doi:10.1063/1.445017.

Flood H (1934). Tröpfchenbildung in übersättigten Äthylalkohol-Wasserdampfgemischen [Formation of droplets in supersaturated mixtures of ethyl alcohol and water vapor]. *Zeitschrift für Physikalische Chemie* **170A**, 286–294. doi:10.1515/zpch-1934-17026.

Ford IJ (1997). Nucleation theorems, the statistical mechanics of molecular clusters, and a revision of classical nucleation theory. *Physical Review E* **56**, 5615–5629. doi:10.1103/PhysRevE.56.5615.

Ford IJ and Clement CF (1989). The effects of temperature fluctuations in homogeneous nucleation theory. *Journal of Physics A: Mathematical and General.* **22**, 4007–4018. doi:10.1088/0305-4470/22/18/033.

Frenklach M (2002). Reaction mechanism of soot formation in flames. *Physical Chemistry Chemical Physics* **4**, 2028–2037. doi:10.1039/b110045a.

Frenklach M and Wang H (1991). Detailed modeling of soot particle nucleation and growth. *Symposium (International) on Combustion* **23**, 1559–1566. doi:10.1016/S0082-0784(06)80426-1.

Fridman AA, Boufendi L, Hbid T, Potapkin BV, and Bouchoule A (1996). Dusty plasma formation: Physics and critical phenomena. Theoretical approach. *Journal of Applied Physics* **79**, 1303–1314. 10.1063/1.361026.

Friedlander SK (1983). Dynamics of aerosol formation by chemical reaction. *Annals of the New York Academy of Sciences* **404**, 354–364. doi:10.1111/j.1749–6632.1983.tb19497.x.

Friedlander SK (2000). *Smoke, Dust and Haze: Fundamentals of Aerosol Dynamics*. New York: Oxford University Press.

Gai H, Thompson DL, and Raff LM (1988). Trajectory study of the formation and decay of silicon trimer complexes in monomer-dimer collisions. *Journal of Chemical Physics* **88**, 156–162. doi:10.1063/1.454647.

Gallagher A, Howling AA, and Hollenstein C (2002). Anion reactions in silane plasma. *Journal of Applied Physics* **91**, 5571–5580. doi:10.1063/1.1459758.

Gelbard F and Seinfeld JH (1979). The general dynamic equation for aerosols: Theory and application to aerosol formation and growth. *Journal of Colloid and Interface Science* **68**, 363–382. doi:10.1016/0021-9797(79)90289-3.

Gelbard F, Tambour Y, and Seinfeld JH (1980). Sectional representations for simulating aerosol dynamics. *Journal of Colloid and Interface Science* **76**, 541–556. doi:10.1016/0021-9797(80)90394-X.

Ghosh D, Manka A, Strey R, Seifert S, Winans RE, and Wyslouzil BE (2008). Using small angle x-ray scattering to measure the homogeneous nucleation rates of n-propanol, n-butanol, and n-pentanol in supersonic nozzle expansions. *Journal of Chemical Physics* **129**, 124302. doi:10.1063/1.2978384.

Girshick SL (1997). Theory of nucleation from the gas phase by a sequence of reversible chemical reactions. *Journal of Chemical Physics* **107**, 1948–1952. doi:10.1063/1.475050.

Girshick SL (2014). The dependence of homogeneous nucleation rate on supersaturation. *Journal of Chemical Physics* **141**, 024307. doi:10.1063/1.4887338.

Girshick SL and Chiu C-P (1990). Kinetic nucleation theory: A new expression for the rate of homogeneous nucleation from an ideal supersaturated vapor. *Journal of Chemical Physics* **93**, 1273–1277. doi:10.1063/1.459191.

Girshick SL, Chiu C-P, and McMurry PH (1990). Time-dependent aerosol models and homogeneous nucleation rates. *Aerosol Science and Technology* **13**, 465–477. doi:10.1080/02786829008959461.

Girshick SL, Swihart MT, Nijhawan S, Suh S-M, and Mahajan MR (2000). Numerical modeling of gas-phase nucleation and particle growth during chemical vapor deposition of silicon. *Journal of the Electrochemical Society* **147**, 2303–2311. doi:10.1149/1.1393525.

Girshick SL, Agarwal P, and Truhlar DG (2009). Homogeneous nucleation with magic numbers: Aluminum. *Journal of Chemical Physics* **131**, 134305. doi:10.1063/1.3239469.

Giunta CJ, McCurdy RJ, Chapple-Sokol JD, and Gordon RG (1990). Gas-phase kinetics in the atmospheric pressure chemical vapor deposition of silicon from silane and disilane. *Journal of Applied Physics* **67**, 1062–1075. doi:10.1063/1.345792.

Goldstein AN, Echer CM, and Alivisatos AP (1992). Melting in semiconductor nanocrystals. *Science* **256**, 1425–1427. doi:10.1126/science.256.5062.1425.

Grassmann A and Peters F (2000). Homogeneous nucleation rates of n-pentanol in nitrogen measured in a piston-expansion tube. *Journal of Chemical Physics* **113**, 6774–6781. doi:10.1063/1.1310597.

Grassmann A and Peters F (2002). Homogeneous nucleation rates of n-propanol in nitrogen measured in a piston-expansion tube. *Journal of Chemical Physics* **116**, 7617–7620. doi:10.1063/1.1465400.

Grinin AP and Kuni FM (1989). Thermal and fluctuation effects of nonisothermal nucleation. *Theoretical and Mathematical Physics* **80**, 968–980. doi:10.1007/BF01016191.

Haar L, Gallagher JS and Kell GS (1984). *NBS/NRC Steam Tables: Thermodynamic and Transport Properties and Computer Programs for Vapor and Liquid States of Water in SI Units*. New York: Hemisphere Publishing Company.

Hamill P, Cadle RD, and Kiang CS (1977). The nucleation of H_2SO_4-H_2O solution aerosol particles in the stratosphere. *Journal of Atmospheric Sciences* **34**, 150–162. doi:10.1175/1520-0469(1977)034 < 0150:Tnohhs>http://2.0.Co;2.

Haye MJ and Bruin C (1994). Molecular dynamics study of the curvature correction to the surface tension. *Journal of Chemical Physics* **100**, 556–559. doi:10.1063/1.466972.

Heath CH (2001). Binary condensation in a supersonic nozzle. Ph.D. dissertation, Department of Chemical Engineering, Worcester Polytechnic University, Worcester, MA.

Heath CH, Streletzky K, Wyslouzil BE, Wölk J, and Strey R (2002). H_2O-D_2O condensation in a supersonic nozzle. *Journal of Chemical Physics* **117**, 6176–6185. doi:10.1063/1.1502644.

Heist RH and He H (1994). Review of vapor to liquid homogeneous nucleation experiments from 1968 to 1992. *Journal of Physical and Chemical Reference Data* **23**, 781–805. doi:10.1063/1.555951.

Heist RH and Reiss H (1973). Investigation of the homogeneous nucleation of water vapor using a diffusion cloud chamber. *Journal of Chemical Physics* **59**, 665–671. doi:10.1063/1.1680073.

Hinds WC (1999). *Aerosol Technology: Properties, Behavior, and Measurement of Airborne Particles*. New York: Wiley.

Holten V, Labetski DG, and van Dongen MEH (2005). Homogeneous nucleation of water between 200 and 240 K: New wave tube data and estimation of the Tolman length. *Journal of Chemical Physics* **123**, 104505. doi:10.1063/1.2018638.

Howling AA, Dorier J-L and Hollenstein C (1993a). Negative ion mass spectra and particulate formation in radio frequency silane plasma deposition experiments. *Applied Physics Letters* **62**, 1341–1343. doi:10.1063/1.108724.

Howling AA, Sansonnens L, Dorier J-L and Hollenstein C (1993b). Negative hydrogenated silicon ion clusters as particle precursors in RF silane plasma deposition experiments. *Journal of Physics D: Applied Physics* **26**, 1003–1006. doi:10.1088/0022-3727/26/6/019.

Howling AA, Sansonnens L, Dorier J-L, and Hollenstein C (1994). Time-resolved measurements of highly polymerized negative ions in radio frequency silane plasma deposition experiments. *Journal of Applied Physics* **75**, 1340–1353. doi:10.1063/1.356413.

Howling AA, Courteille C, Dorier J-L, Sansonnens L, and Hollenstein C (1996). From molecules to particles in silane plasmas. *Pure and Applied Chemistry* **68**, 1017–1022. doi:10.1351/pac199668051017.

Hruby J, Viisanen Y, and Strey R (1996). Homogeneous nucleation rates for *n*-pentanol in argon: Determination of the critical cluster size. *Journal of Chemical Physics* **104**, 5181–5187. doi:10.1063/1.471145.

Hultgren R (1973). *Selected Values of the Thermodynamic Properties of the Elements*. Metals Park, OH: American Society for Metals.

Hung C-H, Krasnopoler MJ, and Katz JL (1989). Condensation of a supersaturated vapor. VIII. The homogeneous nucleation of *n*-nonane. *Journal of Chemical Physics* **90**, 1856–1865. doi:10.1063/1.456027.

Iland K, Wedekind J, Wölk J, Wagner PE, and Strey R (2004). Homogeneous nucleation rates of 1-pentanol. *Journal of Chemical Physics* **121**, 12259–12264. doi:10.1063/1.1809115.

Iland K, Wölk J, Strey R, and Kashchiev D (2007). Argon nucleation in a cryogenic nucleation pulse chamber. *Journal of Chemical Physics* **127**, 154506. doi:10.1063/1.2764486.

Kacker A and Heist RH (1985). Homogeneous nucleation rate measurements. 1. Ethanol, *n*-propanol, and *i*-propanol. *Journal of Chemical Physics* **82**, 2734–2744. doi:10.1063/1.448271.

Kalikmanov VI (2006). Mean-field kinetic nucleation theory. *Journal of Chemical Physics* **124**, 124505. doi:10.1063/1.2178812.

Kantrowitz A (1951). Nucleation in very rapid vapor expansions. *Journal of Chemical Physics* **19**, 1097–1100.

Kashchiev D (1982). On the relation between nucleation work, nucleus size, and nucleation rate. *Journal of Chemical Physics* **76**, 5098–5102. doi:10.1063/1.442808.

Kashchiev D (2000). *Nucleation: Basic Theory with Applications*. Oxford: Butterworth Heinemann.

Katz JL and Donohue MD (1982). Nucleation with simultaneous chemical reaction. *Journal of Colloid and Interface Science* **85**, 267–277. doi:10.1016/0021-9797(82)90255-7.

Katzer G, Ernst MC, Sax AF, and Kalcher J (1997). Computational thermochemistry of medium-sized silicon hydrides. *Journal of Physical Chemistry A* **101**, 3942–3958. doi:10.1021/jp9631947.

Kelkar M, Rao NP, and Girshick SL (1996). Homogeneous nucleation of silicon: Effects of the properties and kinetics of small structured clusters. In Kulmala M and Wagner PE (eds), *Nucleation and Atmospheric Aerosols 1996*. Oxford: Elsevier Science Ltd, 117–120.

Keshavarz F, Kubečka J, Attoui M, Vehkamäki H, Kurtén T, and Kangasluoma J (2020). Molecular origin of the sign preference of ion- induced heterogeneous nucleation in a complex ionic liquid–diethylene glycol system. *Journal of Physical Chemistry C* **124**, 26944–26952. doi:10.1021/acs.jpcc.0c09481.

Kholghy MR, Kelesidis GA, and Pratsinis SE (2018). Reactive polycyclic aromatic hydrocarbon dimerization drives soot nucleation. *Physical Chemistry Chemical Physics* **20**, 10926–10938. doi:10.1039/c7cp07803j.

Kildgaard JV, Mikkelsen KV, Bilde M, and Elm J (2018). Hydration of atmospheric molecular clusters: A new method for systematic configurational sampling. *Journal of Physical Chemistry A* **122**, 5026–5036. doi:10.1021/acs.jpca.8b02758.

Kim YJ, Wyslouzil BE, Wilemski G, Wölk J, and Strey R (2004). Isothermal nucleation rates in supersonic nozzles and the properties of small water clusters. *Journal of Physical Chemistry A* **108**, 4365–4377. doi:10.1021/jp037030j.

Kulmala M, Laaksonen A, and Girshick SL (1992). The self-consistency correction to homogeneous nucleation: Extension to binary systems. *Journal of Aerosol Science* **23**, 309–312. doi:10.1016/0021-8502(92)90331-o.

Laaksonen A (1992). Nucleation of binary water-normal-alcohol vapors. *Journal of Chemical Physics* **97**, 1983–1989. doi:10.1063/1.463136.

Landau L and Teller E (1936). Theory of sound dispersion. *Physik Zeitschrift der Sowjetunion* **10**, 34–38.

Le Picard R, Markosyan AH, Porter DH, Girshick SL, and Kushner MJ (2016). Synthesis of silicon nanoparticles in nonthermal capacitively-coupled flowing plasmas: Processes and transport. *Plasma Chemistry and Plasma Processing* **36**, 941–972. doi:10.1007/s11090-016-9721-6.

Lei YA, Bykov T, Yoo S, and Zeng XC (2005). The Tolman length: Is it positive or negative? *Journal of the American Chemical Society* **127**, 15346–15347. doi:10.1021/ja054297i.

Lemmon EW, Bell IH, Huber ML, and McLinden MO (2023). Thermophysical properties of fluid systems. In Linstrom PJ and Mallard WG (eds), *NIST Chemistry WebBook*, NIST Standard Reference Database Number 69, Gaithersburg, MD: National Institute of Standards and Technology. doi:10.18434/T4D303 (accessed May 12, 2023).

Li ZH and Truhlar DG (2008). Cluster and nanoparticle condensation and evaporation reactions. Thermal rate constants and equilibrium constants for $Al_m + Al_{n-m} = Al_n$ with $n = 2$–60 and $m = 1$–8. *Journal of Physical Chemistry C* **112**, 11109–11121. doi:10.1021/jp711349v.

Li ZH, Bhatt D, Schultz NE, Siepmann JI, and Truhlar DG (2007). Free energies of formation of clusters and nanoparticles from molecular simulations: Al$_n$ with n = 2–60. *Journal of Physical Chemistry C* **111**, 16227–16242. doi:10.1021/jp073559v.

Lighty JS, Veranth JM, and Sarofim AF (2000). Combustion aerosols: Factors governing their size and composition and implications to human health. *Journal of the Air & Waste Management Association* **50**, 1565–1618. doi:10.1080/10473289.2000.10464197.

Lihavainen H, Viisanen Y, and Kulmala M (2001). Homogeneous nucleation of *n*-pentanol in a laminar flow diffusion chamber. *Journal of Chemical Physics* **114**, 10031–10038. doi:10.1063/1.1368131.

Luijten CCM (1998). Nucleation and droplet growth at high pressure. Ph.D. dissertation, Eindhoven University of Technology, Eindhoven, Netherlands.

Luijten CCM, Baas ODE, and vanDongen EH (1997a). Homogeneous nucleation rates for *n*-pentanol from expansion wave tube experiments. *Journal of Chemical Physics* **106**, 4152–4156. doi:10.1063/1.473125.

Luijten CCM, Bosschaart KJ, and vanDongen MEH (1997b). High pressure nucleation in water/nitrogen systems. *Journal of Chemical Physics* **106**, 8116–8123. doi:10.1063/1.473818.

Maheshwary S, Patel N, Sathyamurthy N, Kulkarni AD, and Gadre SR (2001). Structure and stability of water clusters $(H_2O)_n$, *n* = 8–20: An ab initio investigation. *Journal of Physical Chemistry A* **105**, 10525–10537. doi:10.1021/jp013141b.

Manka AA, Brus D, Hyvarinen AP, Lihavainen H, Wölk J, and Strey R (2010). Homogeneous water nucleation in a laminar flow diffusion chamber. *Journal of Chemical Physics* **132**, 244505. doi:10.1063/1.3427537.

Manka AA, Wedekind J, Ghosh D, Hohler K, Wölk J, and Strey R (2012). Nucleation of ethanol, propanol, butanol, and pentanol: A systematic experimental study along the homologous series. *Journal of Chemical Physics* **137**, 054316. doi:10.1063/1.4739096.

Martin DL, Raff LM, and Thompson DL (1990). Silicon dimer formation by three-body recombination. *Journal of Chemical Physics* **92**, 5311–5318. doi:10.1063/1.458602.

McGraw R and Laviolette RA (1995). Fluctuations, temperature, and detailed balance in classical nucleation theory. *Journal of Chemical Physics* **102**, 8983–8994. doi:10.1063/1.468952.

McMurry PH and Friedlander SK (1979). New particle formation in the presence of an aerosol. *Journal of Colloid and Interface Science* **78**, 513–527. doi:10.1016/0004-6981(79)90322-6.

Mikheev VB, Irving PM, Laulainen NS, Barlow SE, and Pervukhin VV (2002). Laboratory measurement of water nucleation using a laminar flow tube reactor. *Journal of Chemical Physics* **116**, 10772–10786. doi:10.1063/1.1480274.

Miller JA and Melius CF (1992). Kinetic and thermodynamic issues in the formation of aromatic-compounds in flames of aliphatic fuels. *Combustion and Flame* **91**, 21–39. doi:10.1016/0010-2180(92)90124-8.

Miller RC (1976). A comprehensive chamber study of homogeneous nucleation of water over a wide range of temperatures and nucleation rates. Ph.D. dissertation, Department of Physics, University of Missouri, Rolla.

Miller RC, Anderson RJ, Kassner JL, and Hagen DE (1983). Homogeneous nucleation rate measurements for water over a wide range of temperature and nucleation rate. *Journal of Chemical Physics* **78**, 3204–3211. doi:10.1063/1.445236.

Mirabel P and Katz JL (1974). Binary homogeneous nucleation as a mechanism for the formation of aerosols. *Journal of Chemical Physics* **60**, 1138–1144. doi:10.1063/1.1681124.

Nadykto AB, Yu F, and Herb J (2008). Towards understanding the sign preference in binary atmospheric nucleation. *Physical Chemistry Chemical Physics* **10**, 7073–7078. doi:10.1039/b807415a.

Napari I, Noppel M, Vehkamäki H, and Kulmala M (2002). An improved model for ternary nucleation of sulfuric acid-ammonia-water. *Journal of Chemical Physics* **116**, 4221–4227. doi:10.1063/1.1450557.

Nijmeijer MJP, Bruin C, Vanwoerkom AB, Bakker AF, and Vanleeuwen JMJ (1992). Molecular dynamics of the surface tension of a drop. *Journal of Chemical Physics* **96**, 565–576. doi:10.1063/1.462495.

Nobel Foundation (1965). In *Physics, 1922–1941*. Amsterdam: Elsevier.

NobelPrize.org (2021). *The Nobel Prize in Physics 1927*. Nobel Prize Outreach AB 2023. www.nobelprize.org/prizes/physics/1927/summary (accessed May 10, 2023).

Onischuk AA, Purtov PA, Baklanov AM, Karasev VV, and Vosel SV (2006). Evaluation of surface tension and Tolman length as a function of droplet radius from experimental nucleation rate and supersaturation ratio: Metal vapor homogeneous nucleation. *Journal of Chemical Physics* **124**, 014506. doi:10.1063/1.2140268.

Oxtoby DW (1992). Homogeneous nucleation: Theory and experiment. *Journal of Physics: Condensed Matter* **4**, 7627–7650. doi:10.1088/0953-8984/4/38/001.

Perrin J, Böhm C, Etemadi R, and Lioret A (1994). Possible routes for cluster growth and particle formation in rf silane discharges. *Plasma Sources Science and Technology* **3**, 252–261. doi:10.1088/0963-0252/3/3/003.

Perrin J, Leroy O, and Bordage MC (1996). Cross-sections, rate constants and transport coefficients in silane plasma chemistry. *Contributions to Plasma Physics* **36**, 3–49. doi:10.1002/ctpp.2150360102.

Pesthy AJ, Flagan RC, and Seinfeld JH (1981). The effect of a growing aerosol on the rate of homogeneous nucleation of a vapor. *Journal of Colloid and Interface Science* **82**, 465–479. doi:10.1016/0021-9797(81)90388-X.

Peters F and Paikert B (1989). Nucleation and growth rates of homogeneously condensing water vapor in argon from shock tube experiments. *Experiments in Fluids* **7**, 521–530.

Preining O (1998). The physical nature of very, very small particles and its impact on their behaviour. *Journal of Aerosol Science* **29**, 481–495. doi:10.1016/S0021-8502(97)10046-5.

Rao NP and McMurry PH (1989). Nucleation and growth of aerosol in chemically reacting systems: A theoretical study of the near collision-controlled regime. *Aerosol Science and Technology* **11**, 120–132. doi:10.1080/02786828908959305.

Rao NP and McMurry PH (1990). Effect of the Tolman surface tension correction on nucleation in chemically reacting systems. *Aerosol Science and Technology* **13**, 183–195. doi:10.1080/02786829008959436.

Rasmussen FR, Kubecka J, Besel V, Vehkamäki H, Mikkelsen KV, Bilde M, and Elm J (2020). Hydration of atmospheric molecular clusters III: Procedure for efficient free energy surface exploration of large hydrated clusters. *Journal of Physical Chemistry A* **124**, 5253–5261. doi:10.1021/acs.jpca.0c02932.

Reguera D, Bowles RK, Djikaev Y, and Reiss H (2003). Phase transitions in systems small enough to be clusters. *Journal of Chemical Physics* **118**, 340–353. doi:10.1063/1.1524192.

Reiss H (1950). The kinetics of phase transitions in binary systems. *Journal of Chemical Physics* **18**, 840–848. doi:10.1063/1.1747784.

Rudek MM, Katz JL, Vidensky IV, Zdimal V, and Smolik J (1999). Homogeneous nucleation rates of *n*-pentanol measured in an upward thermal diffusion cloud chamber. *Journal of Chemical Physics* **111**, 3623–3629. doi:10.1063/1.479642.

Ruscic, B (2013). Active thermochemical tables: Water and water dimer. *Journal of Physical Chemistry* **117**, 11940–11953. doi: 10.1021/jp403197t.

Russell KC (1969). Nucleation on gaseous ions. *Journal of Chemical Physics* **50**, 1809–1816. doi:10.1063/1.1671276.

Salpeter EE (1973). Heat transfer in nucleation theory. *Journal of Chemical Physics* **58**, 4331–4337. doi:10.1063/1.1678990.

Sarou-Kanian V, Millot F, and Rifflet JC (2003). Surface tension and density of oxygen-free liquid aluminum at high temperature. *International Journal of Thermophysics.* **24**, 277–286. doi:10.1023/A:1022466319501.

Schmitt JL and Doster GJ (2002). Homogeneous nucleation of *n*-pentanol measured in an expansion cloud chamber. *Journal of Chemical Physics* **116**, 1976–1978. doi:10.1063/1.1429953.

Schmitt JL, Adams GW, and Zalabsky RA (1982). Homogeneous nucleation of ethanol. *Journal of Chemical Physics* **77**, 2089–2097. doi:10.1063/1.444014.

Seinfeld JH and Pandis SN (1998). *Atmospheric Chemistry and Physics: From Air Pollution to Climate Change.* New York: Wiley.

Shi G, Seinfeld JH, and Okuyama K (1990). Transient kinetics of nucleation. *Physical Review A* **41**, 2101–2108. doi:10.1103/PhysRevA.41.2101.

Shields RM, Temelso B, Archer KA, Morrell TE, and Shields GC (2010). Accurate predictions of water cluster formation, $(H_2O)_{n = 2-10}$. *Journal of Physical Chemistry A* **114**, 11725–11737. doi:10.1021/jp104865w.

Simpson JA and Weiner ESC (eds) (1989). *The Oxford English Dictionary*, 2nd ed. Oxford: Clarendon Press; New York: Oxford University Press.

Sinha S, Bhabhe A, Laksmono H, Wölk J, Strey R, and Wyslouzil B (2010). Argon nucleation in a cryogenic supersonic nozzle. *Journal of Chemical Physics* **132**, 064304. doi:10.1063/1.3299273.

Smooke MD, Hall RJ, Colket MB, Fielding J, Long MB, McEnally CS, and Pfefferle LD (2004). Investigation of the transition from lightly sooting towards heavily sooting co-flow ethylene diffusion flames. *Combustion Theory and Modelling* **8**, 593–606. doi:10.1088/1364-7830/8/3/009.

Stauffer D (1976). Kinetic theory of two-component ("hetero-molecular") nucleation and condensation. *Journal of Aerosol Science* **7**, 319–333. doi:10.1016/0021-8502(76)90086-0.

Strey R, Wagner PE, and Schmeling T (1986). Homogeneous nucleation rates for *n*-alcohol vapors measured in a two-piston expansion chamber. *Journal of Chemical Physics* **84**, 2325–2335. doi:10.1063/1.450396.

Suh SM, Girshick SL, and Zachariah MR (2003). The role of total pressure in gas-phase nucleation: A diffusion effect. *Journal of Chemical Physics* **118**, 736–745. doi:10.1063/1.1490345.

Swihart MT (2000). Electron affinities of selected hydrogenated silicon clusters (Si_xH_y, $x = 1-7$, $y = 0-15$) from density functional theory calculations. *Journal of Physical Chemistry A* **104**, 6083–6087. doi:10.1021/jp000626b.

Swihart MT and Girshick SL (1999). Thermochemistry and kinetics of silicon hydride cluster formation during thermal decomposition of silane. *Journal of Physical Chemistry B* **103**, 64–76. doi:10.1021/jp983358e.

Temelso B, Archer KA, and Shields GC (2011). Benchmark structures and binding energies of small water clusters with anharmonicity corrections. *Journal of Physical Chemistry A* **115**, 12034–12046. doi:10.1021/jp2069489.

ter Horst JH, Bedeaux D, and Kjelstrup S (2011). The role of temperature in nucleation processes. *Journal of Chemical Physics* **134**, 054703. doi:10.1063/1.3544689.

Thomson JJ (1888). *Applications of Dynamics to Physics and Chemistry.* London: Cambridge University Press.

Tohmfor G and Volmer M (1938). Die keimbildung unter dem einfluss elektrischer landungen [Nucleation under the influence of electrical charging]. *Annalen der Physik (Leipzig), Series 5* **33**, 109–131.

Tolman RC (1949). The effect of droplet size on surface tension. *Journal of Chemical Physics* **17**, 333–337. doi:10.1063/1.1747247.

Vehkamäki H (2006). *Classical Nucleation Theory in Multicomponent Systems.* Berlin: Springer.

Viisanen Y and Strey R (1994). Homogeneous nucleation rates for *n*-butanol. *Journal of Chemical Physics* **101**, 7835–7843. doi:10.1063/1.468208.

Viisanen Y, Strey R, and Reiss H (1993). Homogeneous nucleation rates for water. *Journal of Chemical Physics* **99**, 4680–4692. doi:10.1063/1.466066.

Viisanen Y, Strey R, Laaksonen A, and Kulmala M (1994). Measurement of the molecular content of binary nuclei. 2. Use of the nucleation rate surface for water-ethanol. *Journal of Chemical Physics* **100**, 6062–6072. doi:10.1063/1.467117.

Viisanen Y, Wagner PE, and Strey R (1998). Measurement of the molecular content of binary nuclei. IV. Use of the nucleation rate surfaces for the *n*-nonane-*n*-alcohol series. *Journal of Chemical Physics* **108**, 4257–4266. doi:10.1063/1.475825.

Viisanen Y, Strey R, and Reiss H (2000). Erratum: "Homogeneous nucleation rates for water" [Journal of Chemical Physics 99, 4680 (1993)]. *Journal of Chemical Physics* **112**, 8205–8206. doi:10.1063/1.481368.

Vincenti WG and Kruger CH (1965). *Introduction to Physical Gas Dynamics.* New York: Wiley.

Volmer M (1939). *Kinetics of Phase Formation.* Dresden, Germany: Theodor Steinkopff Verlag.

Volmer M and Weber A (1926). Keimbildung in übersättigten Gebilden [Nucleation in supersaturated structures]. *Zeitschrift für Physikalische Chemie (Leipzig)* **119 U**, 277–301. doi:10.1515/zpch-1926-11927.

Wagner PE and Strey R (1981). Homogeneous nucleation rates of water vapor measured in a two-piston expansion chamber. *Journal of Physical Chemistry* **85**, 2694–2698. doi:10.1021/j150618a026.

Wagner PE and Strey R (1984). Measurements of homogeneous nucleation rates for *n*-nonane vapor using a two-piston expansion chamber. *Journal of Chemical Physics* **80**, 5266–5275. doi:10.1063/1.446554.

Wang H and Frenklach M (1997). A detailed kinetic modeling study of aromatics formation in laminar premixed acetylene and ethylene flames. *Combustion and Flame* **110**, 173–221. doi:10.1016/s0010-2180(97)00068-0.

Watanabe Y, Shiratani M, Fukuzawa T, Kawasaki H, Ueda Y, Singh S, and Ohkura H (1996). Contribution of short lifetime radicals to growth of particles in SiH_4 HF discharges and effects of particles on deposited films. *Journal of Vacuum Science and Technology A* **14**, 995–1001. doi:10.1116/1.580069.

Wilemski G and Wyslouzil BE (1995). Binary nucleation kinetics. I. Self-consistent size distribution. *Journal of Chemical Physics* **103**, 1127–1136. doi:10.1063/1.469823.

Wilson CTR (1897). Condensation of water vapour in the presence of dust-free air and other gases. *Philosophical Transactions of the Royal Society of London, Series A* **189**, 265–307. doi:10.1098/rsta.1897.0011.

Wilson CTR (1899). On the condensation nuclei produced in gases by the action of Röntgen rays, uranium rays, ultra-violet light, and other agents. *Philosophical Transactions of the Royal Society of London, Series A* **192**, 403–453. doi:10.1098/rsta.1899.0009.

Wilson CTR (1927). On the cloud method of making visible ions and the tracks of ionizing particles. Nobel Lecture, Dec. 12, 1927. In *Nobel Lectures, Physics, 1922–1941*. Amsterdam: Elsevier, 194–214.

Wölk J and Strey R (2001). Homogeneous nucleation of H_2O and D_2O in comparison: The isotope effect. *Journal of Physical Chemistry B* **105**, 11683–11701. doi:10.1021/jp0115805.

Wölk J, Strey R, Heath CH, and Wyslouzil BE (2002). Empirical function for homogeneous water nucleation rates. *Journal of Chemical Physics* **117**, 4954–4960. doi:10.1063/1.1498465.

Wu JJ and Flagan RC (1988). A discrete-sectional solution to the aerosol dynamic equation. *Journal of Colloid and Interface Science* **123**, 339–352.

Wyslouzil BE and Seinfeld JH (1992). Nonisothermal homogeneous nucleation. *Journal of Chemical Physics* **97**, 2661–2670. doi:10.1063/1.463055.

Wyslouzil BE and Wölk J (2016). Overview: Homogeneous nucleation from the vapor phase – The experimental science. *Journal of Chemical Physics* **145**, 211702. doi:10.1063/1.4962283.

Yang H, Drossinos Y, and Hogan CJ (2019). Excess thermal energy and latent heat in nanocluster collisional growth. *Journal of Chemical Physics* **151**, 224304. doi:10.1063/1.5129918.

Zeldovich JB (1943). On the theory of new phase formation, cavitation. *Acta Physicochimica URSS* **18**, 1–22.

Index